HSC Year 12
MATHEMATICS EXTENSION 2

JIM GREEN | JANET HUNTER
SERIES EDITOR: ROBERT YEN

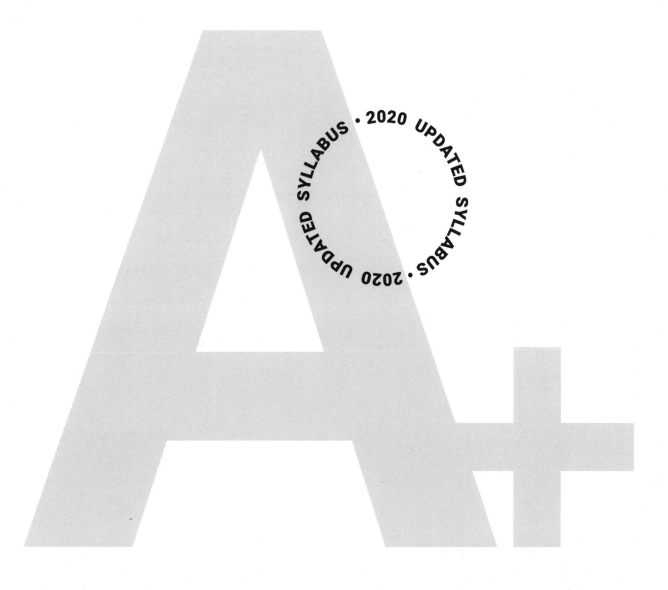

2020 UPDATED SYLLABUS · 2020 UPDATED SYLLABUS · 2020 UPDATED SYLLABUS

+ topic summaries
+ graded practice questions
  with worked solutions
+ HSC exam topic grids (2011–2020)

STUDY NOTES

A+ HSC Mathematics Extension 2 Study Notes
1st Edition
Jim Green
Janet Hunter
ISBN 9780170459266

Publishers: Robert Yen, Kirstie Irwin
Project editor: Tanya Smith
Cover design: Nikita Bansal
Text design: Alba Design
Project designer: Nikita Bansal
Permissions researcher: Corrina Gilbert
Production controller: Karen Young
Typeset by: Nikki M Group Pty Ltd

Any URLs contained in this publication were checked for currency during the production process. Note, however, that the publisher cannot vouch for the ongoing currency of URLs.

NSW Education Standards Authority (NESA) 2016 Higher School Certificate Examination Mathematics Extension 1; NSW Education Standards Authority (NESA) Higher School Certificate Examination Mathematics Extension 2: 2012 – 2014, 2017 - 2020 © NSW Education Standards Authority for and on behalf of the Crown in right of the State of New South Wales.

For product information and technology assistance,
in Australia call **1300 790 853**;
in New Zealand call **0800 449 725**

For permission to use material from this text or product, please email **aust.permissions@cengage.com**

ISBN 978 0 17 045926 6

Cengage Learning Australia
Level 7, 80 Dorcas Street
South Melbourne, Victoria Australia 3205

Cengage Learning New Zealand
Unit 4B Rosedale Office Park
331 Rosedale Road, Albany, North Shore 0632, NZ

For learning solutions, visit **cengage.com.au**

Printed in China by 1010 Printing International Limited.
1 2 3 4 5 6 7 25 24 23 22 21

ABOUT THIS BOOK

Introducing *A+ HSC Year 12 Mathematics*, a new series of study guides designed to help students revise the topics of the new HSC maths courses and achieve success in their exams. *A+* is published by Cengage, the educational publisher of *Maths in Focus* and *New Century Maths*.

For each HSC maths course, Cengage has developed a STUDY NOTES book and a PRACTICE EXAMS book. These study guides have been written by experienced teachers who have taught the new courses, some of whom are involved in HSC exam marking and writing. This is the first study guide series to be published after the first HSC exams of the new courses in 2020, so it incorporates the latest changes to the syllabus and exam format.

This book, *A+ HSC Year 12 Mathematics Extension 2 Study Notes,* contains topic summaries and graded practice questions, grouped into 5 broad topics, addressing the outcomes in the Mathematics Extension 2 syllabus. The topic-based structure means that this book can be used for revision after a topic has been covered in the classroom, as well as for course review and preparation for the trial and HSC exams. Each topic chapter includes a review of the main mathematical concepts, and multiple-choice and short-answer questions with worked solutions. Past HSC examination questions have been included to provide students with the opportunity to see how they will be expected to show their mathematical understanding in the exams. An HSC exam topic grid (2011–2020) guides students to where and how each topic has been tested in past HSC exams.

Mathematics Extension 2 topics

1. Proof
2. 3D vectors
3. Complex numbers
4. Further integration
5. Mechanics

This book contains for each topic:

- Concept map (see p. 2 for an example)
- Glossary and digital flashcards (see p. 3 for an example)
- Topic summary, addressing key outcomes of the syllabus (see p. 4 for an example)
- Practice set 1: 20 multiple-choice questions (see p. 15 for an example)
- Practice set 2: 20 short-answer questions (see p. 19 for an example)
- Questions graded by level of difficulty: foundation ●○○, moderate ●●○, complex ●●●
- Worked solutions to both practice sets
- HSC exam topic grid (2011–2020) (see p. 36 for an example)

The companion A+ PRACTICE EXAMS book is written by the same authors, Jim Green and Janet Hunter (see p.viii), who also wrote the popular textbook *Maths in Focus 12 Mathematics Extension 2*. It contains topic exams and practice HSC exam papers, both of which are written and formatted in the style of the HSC exams, with space for students to write answers. Worked solutions are provided, along with the authors' expert comments and advice, including how each exam question is marked. As a special bonus, the worked solutions to the 2020 HSC exam paper have been included.

This A+ STUDY NOTES book will become a staple resource in your study in the lead-up to your final HSC exams. Revisit it throughout Year 12 to ensure that you do not forget key concepts and skills. Good luck!

CONTENTS

PROOF

3D VECTORS

COMPLEX NUMBERS

CHAPTER 4

FURTHER INTEGRATION

CHAPTER 5

MECHANICS

YEAR 12 COURSE OVERVIEW

See each concept map printed in full size at the beginning of each chapter.

PROOF

The language and methods of proof

- If-then (implication)
- Converse
- Equivalence
- Negation
- Contrapositive
- Sets of numbers
- Proof by contradiction
- Proof by counterexample

Proofs involving numbers and inequalities

- Properties of inequalities
- 'Consider the difference'
- The arithmetic mean-geometric mean inequality
- The triangle inequality

Mathematical induction

- Series and sigma notation
- Divisibility
- Inequalities
- Calculus, probability and geometry
- Recursive formulas

COMPLEX NUMBERS

The complex plane and polar form

- The complex plane, Argand diagram
- Polar (modulus-argument) form
- Properties of modulus and argument
- Multiplying and dividing complex numbers
- Powers of complex numbers

Complex numbers

- Cartesian form $z = a + ib$
- $\mathrm{Re}(z)$, $\mathrm{Im}(z)$
- Adding, subtracting, multiplying, dividing
- Complex conjugate, $\overline{z}$
- Realising the denominator
- Reciprocal
- Square root of a complex number

De Moivre's theorem and solving equations

- De Moivre's theorem
- Solving quadratic equations
- Quadratic equations with complex coefficients
- Polynomial equations and conjugate roots

Euler's formula and exponential form

- Euler's formula $e^{i\theta} = \cos\theta + i\sin\theta$
- Exponential form $z = re^{i\theta}$
- Exponential, Cartesian, polar forms
- Powers of complex numbers in exponential form

Roots, curves and regions

- Roots of unity: location on the unit circle
- Roots of a complex number
- Graphing lines and curves in the complex plane: equations involving modulus and argument
- Graphing regions in the complex plane: inequalities

Complex numbers as vectors

- Adding and subtracting complex numbers
- Multiplying complex numbers
- Conjugates, multiplying by a scalar, multiplying by i

3D VECTORS

Operations with vectors

- 3D vectors
- Unit vectors
- Addition and subtraction
- Multiplication by a scalar
- Magnitude of a vector
- Scalar (dot) product

Geometrical proofs in 2D and 3D

Vector equations of curves

- 3D space and coordinates
- Parametric equations of curves
- Equation of a sphere

Vector equations of lines

- Vector equation of a line $\underline{r} = \underline{a} + \lambda\underline{b}$
- Testing whether a point lies on a line
- Parallel and perpendicular lines

FURTHER INTEGRATION

Integration by substitution

- Let $u = \ldots$
- The substitution may not be given
- Change limits of definite integrals
- Trigonometric substitutions, including t-formulas

Integration by parts

- $\int uv'dx = uv - \int vu'dx$
- Choosing u and v'

Rational functions and partial fractions

- Quadratic denominators that require completing the square
- Linear and quadratic denominators
- Partial fractions: equating coefficients vs substitution
- Often involve logarithmic or inverse trigonometric functions

Integration by recurrence relations

- Integrals I_n, I_{n-1} and I_0
- Usually involves integration by parts

MECHANICS

Simple harmonic motion

- Velocity and acceleration as functions of displacement
- Equations and graphs for simple harmonic motion
- Acceleration, velocity and displacement
- Amplitude, period, phase shift, centre
- Properties of simple harmonic motion: maximum and zero velocity and acceleration

Resisted motion

- Resisted horizontal motion
- $R = kv$ or kv^2
- $v = f(t)$ and $v = f(x)$
- Resisted vertical motion under gravity and other forces
- Terminal velocity

Modelling motion

- Newton's laws of motion: $F = m\ddot{x}$
- Forces and resolving forces
- Force diagrams
- Coefficient of friction

Projectiles and resisted motion

- Equation of the path of a projectile
- Projectile motion under gravity and other forces

SYLLABUS REFERENCE GRID

Topic and subtopics	*A+ HSC Year 12 Mathematics Extension 2 Study Notes* chapter
PROOF	
MEX-P1 The nature of proof	1 Proof
MEX-P2 Further proof by mathematical induction	1 Proof
VECTORS	
MEX-V1 Further work with vectors	2 3D vectors
V1.1 Introduction to three-dimensional vectors	
V1.2 Further operations with three-dimensional vectors	
V1.3 Vectors and vector equations of lines	
COMPLEX NUMBERS	
MEX-N1 Introduction to complex numbers	3 Complex numbers
N1.1 Arithmetic of complex numbers	
N1.2 Geometric representation of a complex number	
N1.3 Other representations of complex numbers	
MEX-N2 Using complex numbers	3 Complex numbers
N2.1 Solving equations with complex numbers	
N2.2 Geometrical implications of complex numbers	
CALCULUS	
MEX-C1 Further integration	4 Further integration
MECHANICS	
MEX-M1 Applications of calculus to mechanics	5 Mechanics
M1.1 Simple harmonic motion	
M1.2 Modelling motion without resistance	
M1.3 Resisted motion	
M1.4 Projectiles and resisted motion	

ABOUT THE AUTHORS

Jim Green was Head of Mathematics at Trinity Catholic College, Lismore, where he spent most of his teaching career of over 35 years. He has written HSC examinations and syllabus drafts, composed questions for the Australian Mathematics Competition and recently co-authored *Maths in Focus 12 Mathematics Extension 2*.

Janet Hunter is Head of Mathematics at Ascham School, Edgecliff, where she has spent most of her teaching career of over 30 years. She has been a senior HSC examiner and judge, an HSC Advice Line adviser, and recently co-authored *Maths in Focus 12 Mathematics Extension 2*.

A+ DIGITAL FLASHCARDS

Revise key terms and concepts online with the A+ Flashcards. Each topic glossary in this book has a corresponding deck of digital flashcards you can use to test your understanding and recall. Just scan the QR code or type the URL into your browser to access them.

Note: You will need to create a free *NelsonNet* account.

https://get.ga/a-hsc-maths-ext-2

9780170459266

HSC EXAM FORMAT

Mathematics Extension 2 students complete two HSC exams: **Mathematics Extension 1** and **Mathematics Extension 2**.

The following information about the exams was correct at the time of printing in 2021. Please check the NESA website in case it has changed. Visit www.educationstandards.nsw.edu.au, select 'Year 11–Year 12', 'Syllabuses A–Z', 'Mathematics Extension 1/Extension 2', then 'Assessment and Reporting'. Scroll down to 'HSC examination specifications'.

Mathematics Extension 1 HSC exam

	Questions	Marks	Recommended time
Section I	10 multiple-choice questions	10	15 min
Section II	4 multi-part short-answer questions, average 15 marks each, including questions worth 4 or 5 marks	60	1 h 45 min
Total		70	2 h

- Reading time: 10 minutes; use this time to preview the whole exam.

- Working time: 2 hours

- Questions focus on Year 12 outcomes but Year 11 knowledge may be examined.

- Answers are to be written in separate answer booklets.

- A reference sheet is provided at the back of the exam paper, and also this book, containing common formulas.

- The 4- or 5-mark questions are usually complex problems that require many steps of working and careful planning.

- Having 2 hours for a total of 70 marks means that you have an average of 1.7 minutes per mark (or approximately 5 minutes for 3 marks).

- If you budget 15 minutes for Section I and 20 minutes per question for Section II, you will then have 25 minutes at the end of the exam to check over your work and/or complete questions you missed.

Mathematics Extension 2 HSC exam

	Questions	Marks	Recommended time
Section I	10 multiple-choice questions	10	15 min
Section II	6 multi-part short-answer questions, average 15 marks each, including questions worth 4 or 5 marks	90	2 h 45 min
Total		100	3 h

- Reading time: 10 minutes

- Working time: 3 hours

- Answers are to be written in separate answer booklets.

- Having 3 hours for a total of 100 marks means that you have an average of 1.8 minutes per mark (or approximately 5 minutes for 3 marks).

- If you budget 15 minutes for Section I and 25 minutes per question for Section II, then you will have 15 minutes at the end of the exam to check over your work and/or complete questions you missed.

STUDY AND EXAM ADVICE

A journey of a thousand miles begins with a single step.
Lao Tzu (c. 570–490 BCE), Chinese philosopher

I've always believed that if you put in the work, the results will come.
Michael Jordan (1963–), American basketball player

Four PRACtical steps for maths study

1. **P**ractise your maths

- Do your homework.
- Learning maths is about mastering a collection of skills.
- You become successful at maths by doing it more, through regular practice and learning.
- Aim to achieve a high level of understanding.

2. **R**ewrite your maths

- Homework and study are not the same thing. Study is your private 'revision' work for strengthening your understanding of a subject.
- Before you begin any questions, make sure you have a thorough understanding of the topic.
- Take ownership of your maths. Rewrite the theory and examples in your own words.
- Summarise each topic to see the 'whole picture' and know it all.

3. **A**ttack your maths

- All maths knowledge is interconnected. If you don't understand one topic fully, then you may have trouble learning another topic.
- Mathematics is not an HSC course you can learn 'by halves' – you have to know it all!
- Fill in any gaps in your mathematical knowledge to see the 'whole picture'.
- Identify your areas of weakness and work on them.
- Spend most of your study time on the topics you find difficult.

4. **C**heck your maths

- After you have mastered a maths skill, such as graphing a quadratic equation, no further learning or reading is needed, just more practice.
- Compared to other subjects, the types of questions asked in maths exams are conventional and predictable.
- Test your understanding with revision exercises, practice papers and past exam papers.
- Develop your exam technique and problem-solving skills.
- Go back to steps 1–3 to improve your study habits.

Topic summaries and concept maps

Summarise each topic when you have completed it, to create useful study notes for revising the course, especially before exams. Use a notebook or folder to list the important ideas, formulas, terminology and skills for each topic. This book is a good study guide, but educational research shows that effective learning takes place when you rewrite learned knowledge in your own words.

A good topic summary runs for 2 to 4 pages. It is a condensed, personalised version of your course notes. This is your interpretation of a topic, so include your own comments, symbols, diagrams, observations and reminders. Highlight important facts using boxes and include a glossary of key words and phrases.

A concept map or mind map is a topic summary in graphic form, with boxes, branches and arrows showing the connections between the main ideas of the topic. This book contains examples of concept maps. The topic name is central to the map, with key concepts or subheadings listing important details and formulas. Concept maps are powerful because they present an overview of a topic on one large sheet of paper. Visual learners absorb and recall information better when they use concept maps.

When compiling a topic summary, use your class notes, your textbook and this study guide. Ask your teacher for a copy of the course syllabus or the school's teaching program, which includes the objectives and outcomes of every topic in dot point form.

Attacking your weak areas

Most of your study time should be spent on attacking your weak areas to fill in any gaps in your maths knowledge. Don't spend too much time on work you already know well, unless you need a confidence boost! Ask your teacher, use this book or your textbook to improve the understanding of your weak areas and to practise maths skills. Use your topic summaries for general revision, but spend longer study periods on overcoming any difficulties in your mastery of the course.

Practising with past exam papers

Why is practising with past exam papers such an effective study technique? It allows you to become familiar with the format, style and level of difficulty expected in an HSC exam, as well as the common topic areas tested. Knowing what to expect helps alleviate exam anxiety. Remember, mathematics is a subject in which the exam questions are fairly predictable. The exam writers are not going to ask too many unusual questions. By the time you have worked through many past exam papers, this year's HSC exam won't seem that much different.

Don't throw your old exam papers away. Use them to identify your mistakes and weak areas for further study. Revising topics and then working on mixed questions is a great way to study maths. You might like to complete a past HSC exam paper under timed conditions to improve your exam technique.

Past HSC exam papers are available at the NESA website: visit www.educationstandards.nsw.edu.au and select 'Year 11 – Year 12', 'HSC exam papers'. NESA marking feedback and guidelines can also be viewed there. Cengage has also published *A+ HSC Year 12 Mathematics Extension 2 Practice Exams*, containing topic exams and practice HSC exam papers. You can find past HSC exam papers with solutions online, in bookstores, at the Mathematical Association of NSW (www.mansw.nsw.edu.au) and at your school (ask your teacher) or library.

Preparing for an exam

- Make a study plan early; don't leave it until the last minute.
- Read and revise your topic summaries.
- Work on your weak areas and learn from your mistakes.
- Don't spend too much time studying concepts you know already.
- Revise by completing revision exercises and past exam papers or assignments.
- Vary the way you study so that you don't become bored: ask someone to quiz you, voice-record your summary, design a poster or concept map, or explain a concept to someone.
- Anticipate the exam:
  - How many questions will there be?
  - What are the types of questions: multiple-choice, short-answer, long-answer, problem-solving?
  - Which topics will be tested?
  - How many marks are there in each section?
  - How long is the exam?
  - How much time should I spend on each question/section?
  - Which formulas are on the reference sheet and how do I use them in the exam?

During an exam

1. Bring all of your equipment, including a ruler and calculator (check that your calculator works and is in RADIANS mode for trigonometric functions and DEGREES for trigonometric measurements). A highlighter pen may help for tables, graphs and diagrams.

2. Don't worry if you feel nervous before an exam – this is normal and can help you to perform better; however, being too casual or too anxious can harm your performance. Just before the exam begins, take deep, slow breaths to reduce any stress.

3. Write clearly and neatly in black or blue pen, not red. Use a pencil only for diagrams and constructions.

4. Use the **reading time** to browse through the exam to see the work that is ahead of you and the marks allocated to each question. Doing this will ensure you won't miss any questions or pages. Note the harder questions and allow more time for working on them. Leave them if you get stuck, and come back to them later.

5. Attempt every question. It is better to do most of every question and score some marks, rather than ignore questions completely and score 0 for them. Don't leave multiple-choice questions unanswered! Even if you guess, you have a chance of being correct.

6. Easier questions are usually at the beginning, with harder ones at the end. Do an easy question first to boost your confidence. Some students like to leave multiple-choice questions until last so that, if they run out of time, they can make quick guesses. However, some multiple-choice questions can be quite difficult.

7. Read each question and identify what needs to be found and what topic/skill it is testing. The number of marks indicates how much time and working out is required. Highlight any important keywords or clues. Do you need to use the answer to the previous part of the question?

8. After reading each question, and before you start writing, spend a few moments planning and thinking.

9. You don't need to be writing all of the time. What you are writing may be wrong and a waste of time. Spend some time considering the best approach.

10. Make sure each answer seems reasonable and realistic, especially if it involves money or measurement.

11. Show all necessary working, write clearly, draw big diagrams, and set your working out neatly. Write solutions to each part underneath the previous step so that your working out goes down the page, not across.

12. Use a ruler to draw (or read) half-page graphs with labels and axes marked, or to measure scale diagrams.

13. Don't spend too much time on one question. Keep an eye on the time.

14. Make sure you have answered the question. Did you remember to round the answer and/or include units? Did you use all of the relevant information given?

15. If a hard question is taking too long, don't get bogged down. If you're getting nowhere, retrace your steps, start again, or skip the question (circle it) and return to it later with a clearer mind.

16. If you make a mistake, cross it out with a neat line. Don't scribble over it completely or use correction fluid or tape (which is time-consuming and messy). You may still score marks for crossed-out work if it is correct, but don't leave multiple answers! Keep track of your answer booklets and ask for more writing paper if needed.

17. Don't cross out or change an answer too quickly. Research shows that often your first answer is the correct one.

18. Don't round your answer in the middle of a calculation. Round at the end only.

19. Be prepared to write words and sentences in your answers, but don't use abbreviations that you've just made up. Use correct terminology and write one or two sentences for 2 or 3 marks, not mini-essays.

20. If you have time at the end of the exam, double-check your answers, especially the more difficult questions or questions you are uncertain about.

Ten exam habits of the best HSC students

1. Has clear and careful working and checks their answers

2. Has a strong understanding of basic algebra and calculation

3. Reads (and answers) the whole question

4. Chooses the simplest and quickest method

5. Checks that their answer makes sense or sounds reasonable

6. Draws big, clear diagrams with details and labels

7. Uses a ruler for drawing, measuring and reading graphs

8. Can explain answers in words when needed, in one or two clear sentences

9. Uses the previous part/s of a question to solve the next part of the question

10. Rounds answers at the end, not before.

Further resources

Visit the NESA website, www.educationstandards.nsw.edu.au, for the following resources.
Select 'Year 11 – Year 12' and then 'Syllabuses A–Z' or 'HSC exam papers'.

• Mathematics Advanced, Extension 1 and Extension 2 syllabuses

• Past HSC exam papers, including marking feedback and guidelines

• Sample HSC questions/exam papers and marking guidelines

Before 2020, 'Mathematics Advanced' was called 'Mathematics' and although 'Mathematics Extension 1/Extension 2' had the same names, they were different courses with some topics that no longer exist. For these exam papers, select 'Year 11 – Year 12', 'Resources archive', 'HSC exam papers archive'.

MATHEMATICAL VERBS

A glossary of 'doing words' common in maths problems and HSC exams

analyse
study in detail the parts of a situation

apply
use knowledge or a procedure in a given situation

calculate
See **evaluate**

classify/identify
state the type, name or feature of an item or situation

comment
express an observation or opinion about a result

compare
show how two or more things are similar or different

complete
fill in detail to make a statement, diagram or table correct or finished

construct
draw an accurate diagram

convert
change from one form to another, for example, from a fraction to a decimal, or from kilograms to grams

decrease
make smaller

describe
state the features of a situation

estimate
make an educated guess for a number, measurement or solution, to find roughly or approximately

evaluate/calculate
find the value of a numerical expression, for example, 3×8^2 or $4x + 1$ when $x = 5$

expand
remove brackets in an algebraic expression, for example, expanding $3(2y + 1)$ gives $6y + 3$

explain
describe why or how

give reasons
show the rules or thinking used when solving a problem. *See also* **justify**

graph
display on a number line, number plane or statistical graph

hence find/prove
calculate an answer or demonstrate a result using previous answers or information supplied

identify
See **classify**

increase
make larger

interpret
find meaning in a mathematical result

justify
give reasons or evidence to support your argument or conclusion. *See also* **give reasons**

measure
determine the size of something, for example, using a ruler to determine the length of a pen

prove
See **show/prove that**

recall
remember and state

show/prove that
(in questions where the answer is given) use calculation, procedure or reasoning to demonstrate that an answer or result is true

simplify
express a result such as a ratio or algebraic expression in its most basic, shortest, neatest form

sketch
draw a rough diagram that shows the general shape or ideas (less accurate than **construct**)

solve
find the value(s) of an unknown pronumeral in an equation or inequality

state
See **write**

substitute
replace part of an expression with another, equivalent expression.

verify
check that a solution or result is correct, usually by substituting back into an equation or referring back to the problem

write/state
give an answer, formula or result without showing any working or explanation (This usually means that the answer can be found mentally, or in one step)

SYMBOLS AND ABBREVIATIONS

Symbol	Meaning	Symbol	Meaning
$<$	is less than	$\therefore$	therefore
$>$	is greater than	$[a, b], a \le x \le b$	the interval of x-values from a to b (including a and b)
$\le$	is less than or equal to	$(a, b), a < x < b$	the interval of x-values between a and b (excluding a and b)
$\ge$	is greater than or equal to		
( )	parentheses, round brackets	$P(E)$	the probability of event E occurring
[ ]	(square) brackets		
{ }	braces	$P(\bar{E})$	the probability of event E not occurring
$\pm$	plus or minus		
π	pi = 3.14159 …	$A \cup B$	A union B, A or B
$\equiv$	is congruent/identical to	$A \cap B$	A intersection B, A and B
$\circ$	degree	$P(A \mid B)$	the probability of A given B
$\angle$	angle	$n!$	n factorial, $n(n-1)(n-2) \dots \times 1$
Δ	triangle, the discriminant		
$\parallel$	is parallel to	$^nC_r, \binom{n}{r}$	the number of combinations of r objects from n objects
$\perp$	is perpendicular to		
x^2	x squared, $x \times x$	^nP_r	the number of permutations of r objects from n objects
x^3	x cubed, $x \times x \times x$		
$\mathbb{N}$	the set of natural numbers	PDF	probability density function
$\mathbb{Z}$	the set of integers	CDF	cumulative distribution function
$\mathbb{Q}$	the set of rational numbers	$X \sim \text{Bin}(n, p)$	X is a random variable of the binomial distribution
$\mathbb{R}$	the set of real numbers		
$\mathbb{C}$	the set of complex numbers	$\hat{p}$	sample proportion
$\in$	is an element of, belongs to	LHS	left-hand side
:	such that	RHS	right-hand side
$\cup$	union	p.a.	per annum (per year)
$\cap$	intersection	cos	cosine ratio
∞	infinity	sin	sine ratio
$|x|$	absolute value or magnitude of x	tan	tangent ratio
$\bar{z}$	the conjugate of z	$\bar{x}$	the mean
$\underset{\sim}{v}$	the vector v	$\mu = E(X)$	the population mean, expected value
$\overrightarrow{AB}$	the vector AB		
$\underset{\sim}{u} \cdot \underset{\sim}{v}$	the scalar product of $\underset{\sim}{u}$ and $\underset{\sim}{v}$	σ	the standard deviation
$\text{proj}_{\underset{\sim}{u}} \underset{\sim}{v}$	the projection of $\underset{\sim}{v}$ onto $\underset{\sim}{u}$	$\text{Var}(X) = \sigma^2$	the variance
$\lim_{h \to 0}$	the limit as $h \to 0$	Q_1	first quartile or lower quartile
$\dfrac{dy}{dx}, y', f'(x)$	the first derivative of y, $f(x)$	Q_2	median (second quartile)
		Q_3	third quartile or upper quartile
$\dfrac{d^2y}{dx^2}, y'', f''(x)$	the second derivative of y, $f(x)$	IQR	interquartile range
		α	alpha
$\dot{x}, v$	$\dfrac{dx}{dt}$, velocity	θ	theta
		m	gradient
$\ddot{x}, a$	$\dfrac{d^2x}{dt^2}$, acceleration	$\forall$	for all
		$\exists$	there exists
$\int f(x)\,dx$	the integral of $f(x)$	$P \Rightarrow Q$	if P then Q, P implies Q
$f^{-1}(x)$	the inverse function of $f(x)$	iff	if and only if
$\sin^{-1}, \arcsin$	the inverse sine function	$P \Leftrightarrow Q$	P if and only if Q
Σ	sigma, the sum of	$\neg P$	not P
$\displaystyle\sum_{r=1}^{n} T_r$	the sum of T_r from $r = 1$ to n	RTP	required to prove
		QED	demonstrated as required

9780170459266

A+ HSC YEAR 12 MATHEMATICS

STUDY NOTES

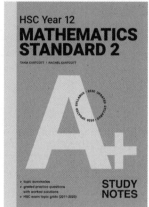

Authors:

Tania Eastcott
Rachel Eastcott

Sarah Hamper

Karen Man
Ashleigh Della Marta

Jim Green
Janet Hunter

PRACTICE EXAMS

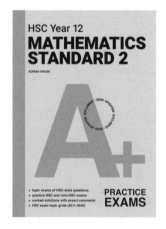

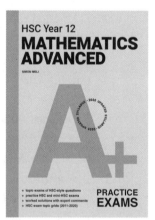

Authors:

Adrian Kruse

Simon Meli

John Drake

Jim Green
Janet Hunter

Jim Green and **Janet Hunter** also wrote the textbook *Maths in Focus 12 Mathematics Extension 2*.

CHAPTER 1
PROOF

9780170459266

PROOF

The language and methods of proof

- If-then (implication)
- Converse
- Equivalence
- Negation
- Contrapositive
- Sets of numbers
- Proof by contradiction
- Proof by counterexample

Proofs involving numbers and inequalities

- Properties of inequalities
- 'Consider the difference'
- The arithmetic mean-geometric mean inequality
- The triangle inequality

Mathematical induction

- Series and sigma notation
- Divisibility
- Inequalities
- Calculus, probability and geometry
- Recursive formulas

Glossary

Mathematical symbols	
$\forall$	for all
$\exists$	there exists
$\in$	is an element of, belongs to
:	such that
$\mathbb{N}$	the set of natural numbers
$\mathbb{Z}$	the set of integers
$\mathbb{Q}$	the set of rational numbers
$\mathbb{R}$	the set of real numbers
$\mathbb{C}$	the set of complex numbers

A+ DIGITAL FLASHCARDS
Revise this topic's key terms and concepts by scanning the QR code or typing the URL into your browser.

https://get.ga/a-hsc-maths-ext-2

contrapositive
The contrapositive statement to 'If P, then Q' ($P \Rightarrow Q$) is 'If not Q, then not P' ($\neg Q \Rightarrow \neg P$). If $P \Rightarrow Q$ is true, then the contrapositive statement $\neg Q \Rightarrow \neg P$ is also true.

converse
The converse to 'If P, then Q' ($P \Rightarrow Q$) is 'If Q, then P' ($Q \Rightarrow P$).

counterexample
An example that shows that a statement is not true, which disproves the statement.

divisibility
The feature of a whole number that allows it to be divisible by another whole number (such as 3) with no remainder.

equivalence
The quality of two statements meaning the same. 'P if and only if Q', also written P iff Q or $P \Leftrightarrow Q$. If $P \Rightarrow Q$ *and* $Q \Rightarrow P$, then we can write $P \Leftrightarrow Q$.

iff
'if and only if'

implication or if-then statement
The quality of one statement implying another statement. 'If P then Q', P implies Q, $P \Rightarrow Q$.

mathematical induction
A method of proving a statement or theorem true for an infinite set of integers by using algebra to generalise from a specific case, such as $n = 1$. Also called **inductive proof** or **proof by induction.**

negation
The 'negative' or 'opposite' statement. The negation of statement P is '*not P*', written $\neg P$, P' or $\overline{P}$.

proof by contradiction
A proof that assumes that the opposite is true and then a logical argument is used to show that it can't be true, therefore the original statement is true.

QED
'*quod erat demonstrandum*', Latin for 'demonstrated as required.'

recursive formula
A formula for calculating the next terms of a sequence based on the previous terms.

RTP
'Required to prove', an abbreviation often used at the start of a proof.

series
A sum of terms of a sequence.
$$S_n = T_1 + T_2 + T_3 + \cdots + T_n$$

sigma notation
A shorthand way of writing a series using the Greek letter sigma Σ.
$$\sum_{r=1}^{n} T_r = T_1 + T_2 + T_3 + \cdots + T_n$$

statement, proposition or premise
A sentence that is true or false but not both. Such statements are usually denoted by a capital letter, for example, we define P as 'It is raining'.

triangle inequality
$|x| + |y| \ge |x + y|$, $\forall x, y \in \mathbb{R}$, which implies geometrically that the sum of the lengths of 2 sides of a triangle is greater than or equal to the length of the third side.

GLOSSARY

Topic summary

The nature of proof (MEX-P1)

Implication or 'if-then' statement

If P and Q stand for the 2 parts of an **implication statement,** we can write the statement 'if P then Q' or 'P implies Q' or '$P \Rightarrow Q$'. An implication statement is also called an **'if-then' statement** or a **conditional statement**.

Pythagoras' theorem is an implication statement: 'If a triangle is right-angled, then the square of the hypotenuse is equal to the sum of the squares of the other 2 sides'.

Another example: 'If the clouds are dark, then it will rain'.

Converse

For the statement 'If P then Q', the **converse** is 'If Q then P'.

The converse of 'P implies Q' $(P \Rightarrow Q)$ is 'Q implies P' $(Q \Rightarrow P)$.

The converse may or may not be true.

> **Hint**
> To find the converse of an if-then statement, swap the 'if' and 'then' parts.

Statement

Pythagoras' theorem: If a triangle is right-angled, then the square of the hypotenuse is equal to the sum of the squares of the other 2 sides.

If the clouds are dark, then it will rain.

Converse

If the square of the longest side of a triangle is equal to the sum of the squares of the 2 shorter sides, then the triangle is right-angled. (True).

If it rains, then the clouds are dark. (Not always true).

Equivalence

If a statement, such as Pythagoras' theorem, and its converse are both true, then we have an **equivalence**.

If $P \Rightarrow Q$ and $Q \Rightarrow P$ then we can write $P \Leftrightarrow Q$, meaning P is equivalent to Q, 'P if and only if Q' or 'P iff Q'.

For example, 'Caitlin is aged 18 or over' and 'Caitlin is legally an adult' are equivalent statements. The if-then statement works both ways:

If Caitlin is aged 18 or over, then she is legally an adult.

If Caitlin is legally an adult, then she is aged 18 or over.

Negation

The **negation** of a statement P is to say '*not P*', written $\neg P$, P' or $\overline{P}$.

If P is true, then $\neg P$ is false. If P is false, then $\neg P$ is true.

Statement

It is hot today.

All boys are tall.

Negation

It is not hot today.

Not all boys are tall, or some boys are not tall.

> **Hint**
> 'The negation of 'all' is 'not all' or 'some did not'.

Note: 'All boys are not tall' or 'All boys are short' is *incorrect*, and is not a negation.

Example 1

What is the negation of this statement?

'For all even numbers n, if n is a multiple of 5, then n is a multiple of 10'.

A n is odd and n is not a multiple of 5 or 10.

B n is even and n is a multiple of 5 but not a multiple of 10.

C If n is even, then n is not a multiple of 5 and n is not a multiple of 10.

D If n is odd, then if n is not a multiple of 5 then n is not a multiple of 10.

Solution

B n is even and n is a multiple of 5 but not a multiple of 10.

Contrapositive

The **contrapositive** statement of 'If P then Q' $(P \Rightarrow Q)$ is 'If not Q then not P' $(\neg Q \Rightarrow \neg P)$.

Statement

Pythagoras' theorem: If a triangle is right-angled, then the square of the hypotenuse is equal to the sum of the squares of the other 2 sides.

If Caitlin is aged 18 or over, then she is legally an adult.

Contrapositive

If the square of the hypotenuse of a triangle is not equal to the sum of the squares of the other 2 sides, then the triangle is not right-angled.

If Caitlin is not legally an adult, then she is not aged 18 or over.

> **Hint**
> To find the contrapositive of an if-then statement, swap the 'if' and 'then' parts (find the converse), then negate each part.

A statement and its contrapositive are equivalent statements. If one is true, then the other is true. If one is false, then the other is false.

> **Hint**
> Sometimes, it is easier to prove that a statement is true by proving that its contrapositive is true.

Example 2   ©NESA   2020 HSC EXAM, QUESTION 15(a)(ii) AND (iii) MODIFIED

In the set of integers, let P be the proposition:

'If $k + 1$ is divisible by 3, then $k^3 + 1$ is divisible by 3'.

a Write down the contrapositive of the proposition P.

b Write down the converse of the proposition P.

Solution

a 'If $k^3 + 1$ is not divisible by 3, then $k + 1$ is not divisible by 3'.

b 'If $k^3 + 1$ is divisible by 3, then $k + 1$ is divisible by 3'.

Sets of numbers

$\mathbb{N}$ the set of natural numbers $\{1, 2, 3, \ldots\}$, though some definitions of natural numbers include 0

$\mathbb{Z}$ the set of integers, including all of $\mathbb{N}$

$\mathbb{Q}$ the set of rational numbers, including all of $\mathbb{Z}$

$\mathbb{R}$ the set of real numbers, including all of $\mathbb{Q}$

$\mathbb{C}$ the set of complex numbers, including all of $\mathbb{R}$

Related symbols

$\forall$ for all

$\exists$ there exists

$\in$ is an element of, belongs to

:   such that

Statement	Meaning
$\forall n \in \mathbb{Z}, \exists M \in \mathbb{Z} : M = 2n$	For all integers n, there exists an integer M such that $M = 2n$.
$\forall r \in \mathbb{Q}, \exists p, q \in \mathbb{Z} : r = \dfrac{p}{q}$	For all rational numbers r, there exist 2 integers p and q such that $r = \dfrac{p}{q}$.

Proof by contradiction

A **proof by contradiction** works by assuming the opposite of a statement (negation) is true and then showing by a logical argument that the assumption is false, so therefore the original statement is true.

Example 3

Prove by contradiction that $\sqrt{2}$ is irrational.

Solution

Assume that $\sqrt{2}$ is rational,

that is, $\exists p, q \in \mathbb{Z}, q \neq 0$ (p and q are integers) such that:

$$\sqrt{2} = \frac{p}{q}$$

where p, q have no common factors.

Then, squaring both sides:

$$2 = \frac{p^2}{q^2}$$

Rearranging:

$$p^2 = 2q^2$$

So p^2 is even, which implies that p is even.

Therefore, we can write $p = 2k$ for some $k \in \mathbb{Z}$.

> **Hint**
> The square of an even number is always even. The square of an odd number is always odd. Why?

So $p^2 = (2k)^2 = 4k^2$ but $p^2 = 2q^2$ also.

$\therefore \ 2q^2 = 4k^2$

$\qquad q^2 = 2k^2$

So q^2 is even, which implies that q is even.

$\therefore \ p$ and q are both even.

Contradiction, as in the assumption p and q have no common factors.

Therefore the original assumption was wrong.

Therefore $\sqrt{2}$ is irrational.   QED.

> **Hint**
>
> The abbreviation **QED** stands for '*quod erat demonstrandum*', which is Latin for 'demonstrated as required.'
>
> It is commonly put at the end of a proof to show that the proof is complete.

Proof by counterexample

A **counterexample** is an example that shows that a statement is not true for all cases.

Example 4

Provide a counterexample to show that $\forall \, n \in \mathbb{R}, \ n^2 + n \geq 0$ is not true.

Solution

The statement says that the square of a real number plus the same number is always positive or zero. It appears to be true for almost all cases. However, if we find one counterexample, then the statement is disproved.

Try a small negative number such as $n = -\dfrac{1}{2}$:

$$
\begin{aligned}
n^2 + n &= \left(-\frac{1}{2}\right)^2 + \left(-\frac{1}{2}\right) \\
&= \frac{1}{4} + \left(-\frac{1}{2}\right) \\
&= -\frac{1}{4} \\
&< 0
\end{aligned}
$$

So this statement is not true.

Example 5   ©NESA   2020 HSC EXAM, QUESTION 7

Consider the proposition:

'If $2^n - 1$ is not prime, then n is not prime'.

Given that each of the following statements is true, which statement disproves the proposition?

A $2^5 - 1$ is prime

B $2^6 - 1$ is divisible by 9

C $2^7 - 1$ is prime

D $2^{11} - 1$ is divisible by 23

Answer

D $2^{11} - 1$ is divisible by 23.

Proofs involving numbers

Example 6   ©NESA   2020 HSC EXAM, QUESTION 15(a)(i)

In the set of integers, let P be the proposition:

'If $k + 1$ is divisible by 3, then $k^3 + 1$ is divisible by 3'.

Prove that the proposition P is true.

Solution

RTP: If $k + 1$ is divisible by 3, then $k^3 + 1$ is divisible by 3.

Let $k + 1 = 3M$ for some positive integer M.

$\therefore k = 3M - 1$

$\begin{aligned}
\therefore k^3 + 1 &= (3M - 1)^3 + 1 \\
&= (3M)^3 - 3(3M)^2 + 3(3M) - 1 + 1 \\
&= 27M^3 - 27M^2 + 9M \\
&= 3(9M^3 - 9M^2 + 3M), \text{ which is divisible by 3.}
\end{aligned}$

OR

Let $k + 1 = 3M$ for some positive integer M.

Consider the factorisation of the sum of 2 cubes:

$\begin{aligned}
k^3 + 1 &= (k + 1)(k^2 - k + 1) \\
&= 3M(k^2 - k + 1), \text{ which is divisible by 3.}
\end{aligned}$

> **Hint**
> RTP = 'Required to prove:'

> **Hint**
> Even though the 'sum and difference of 2 cubes' formula $a^3 \pm b^3 = (a \pm b)(a^2 \mp ab + b^2)$ is no longer part of the Maths Advanced course, it has many applications in the Maths Extension 2 course, in proofs like this and with complex numbers.

Proofs involving inequalities

1. For any 2 real numbers a and b, $a > b$ if $a - b > 0$.

2. For any real number a, $a^2 \geq 0$.

3. If $a^2 \geq b^2$ and $a, b > 0$, then $a \geq b$.

We can use these definitions to prove an inequality, $a > b$, by **considering the difference**, $a - b$, and showing that it is positive ($a - b > 0$) for all cases.

Example 7

a   Prove $\dfrac{p^2 + q^2}{2} \geq pq \; \forall \, p, q \in \mathbb{R}$.

b   Hence, prove $x^2 + y^2 + z^2 \geq xy + yz + zx \; \forall \, x, y, z \in \mathbb{R}$.

Solution

a   Consider the difference:

$\begin{aligned}
\dfrac{p^2 + q^2}{2} - pq &= \dfrac{p^2 + q^2 - 2pq}{2} \\
&= \dfrac{p^2 - 2pq + q^2}{2} \\
&= \dfrac{(p - q)^2}{2}
\end{aligned}$

Now $(p - q)^2 \geq 0 \; \forall \, p, q \in \mathbb{R}$ [equality when $p = q$]

$\therefore \dfrac{(p - q)^2}{2} \geq 0$

$\therefore \dfrac{p^2 + q^2}{2} - pq \geq 0$

So $\dfrac{p^2 + q^2}{2} \geq pq \; \forall \, p, q \in \mathbb{R}$.

b We have just proved that $\dfrac{p^2 + q^2}{2} \geq pq$.

Rearranging, we can also write $p^2 + q^2 \geq 2pq$.

Similarly, $x^2 + y^2 \geq 2xy$
$\quad\qquad y^2 + z^2 \geq 2yz$
$\quad\qquad z^2 + x^2 \geq 2zx$

Add the 3 inequalities together:

$2x^2 + 2y^2 + 2z^2 \geq 2xy + 2yz + 2zx$
$x^2 + y^2 + z^2 \geq xy + yz + zx \;\forall\, x, y, z \in \mathbb{R}$ QED.

The arithmetic mean-geometric mean (AM-GM) inequality

The mean of 2 numbers a and b is $\dfrac{a + b}{2}$. This is also called the **arithmetic mean** because a, $\dfrac{a + b}{2}$ and b form an **arithmetic sequence**.

The **geometric mean** of a and b (where a and b are both positive) is the positive number x such that a, x and b form a **geometric sequence** with common ratio r.

$$\therefore\, r = \frac{x}{a} = \frac{b}{x} \qquad a > 0,\, b > 0,\, x > 0$$

$$x^2 = ab$$
$$x = \sqrt{ab}$$

So the geometric mean of a and b is $\sqrt{ab}$.

The arithmetic mean is always greater than or equal to the geometric mean:

$$\frac{a + b}{2} \geq \sqrt{ab}$$

They are equal when $a = b$.

$$\text{LHS} = \frac{b + b}{2} = \frac{2b}{2} = b$$

$$\text{RHS} = \sqrt{ab} = \sqrt{b^2} = b \qquad (b > 0)$$

The AM-GM inequality is used in many inequality proofs in the Maths Extension 2 course.

Proof:

Consider the difference:

$$\frac{a + b}{2} - \sqrt{ab} = \frac{a + b - 2\sqrt{ab}}{2}$$

$$= \frac{\left(\sqrt{a}\right)^2 - 2\sqrt{ab} + \left(\sqrt{b}\right)^2}{2}$$

$$= \frac{\left(\sqrt{a} - \sqrt{b}\right)^2}{2}$$

$$\frac{\left(\sqrt{a} - \sqrt{b}\right)^2}{2} \geq 0 \text{ as } \left(\sqrt{a} - \sqrt{b}\right)^2 \geq 0$$

$$\therefore\, \frac{a + b}{2} \geq \sqrt{ab}$$

The triangle inequality

The triangle inequality is

$$|x| + |y| \geq |x + y|, \forall x, y \in \mathbb{R},$$

which implies geometrically that the sum of the lengths of 2 sides of a triangle is greater than or equal to the length of the third side.

$$|x| + |y| \geq |z|$$

$$|x| + |y| \geq |x + y|$$

$$|x| + |y| \approx |z|$$

$$|x| + |y| \approx |x + y|$$

If the sum of lengths of the 2 sides are *equal* to the third side, then it's a 'collapsed' (flat) triangle with no area, where the 2 sides form a straight line with the third side.

The triangle inequality is also used in the *3D vectors* and *Complex numbers* topics, where it describes the 2 diagonals of the parallelogram rule.

Further proof by mathematical induction (MEX-P2)

Proof by mathematical induction: steps

1. Show that the statement is true for $n = 1$ (or the smallest value of n given).

2. Assume that the statement is true for $n = k$.

3. Using the assumption, prove that the statement is true for the next integer, $n = k + 1$.

4. Conclusion: 'Hence, by mathematical induction the statement is true for all integers $n \geq 1$'.

Example 8

Prove by mathematical induction that $3^n + 2^n$ is divisible by 5, $\forall n \in \mathbb{N}$, where n is odd.

Solution

Let $P(n)$ be the proposition that $3^n + 2^n = 5B \; \forall n \in \mathbb{N}$, where n is odd and $B \in \mathbb{N}$.

Prove $P(1)$ is true:

$3^1 + 2^1 = 5$, which is a multiple of 5.

So $P(1)$ is true.

Assume $P(k)$ is true for some odd $k \in \mathbb{N}$:

$3^k + 2^k = 5p$ for some $p \in \mathbb{N}$

Make 3^k the subject:

$3^k = 5p - 2^k \quad [*]$

RTP: $P(k + 2)$ is true, $k + 2$ being the next odd number after k, that is,

$3^{k+2} + 2^{k+2} = 5q$ for some $q \in \mathbb{N}$.

Proof:

$$3^{k+2} + 2^{k+2} = 3^2 \times 3^k + 2^2 \times 2^k$$
$$= 9(3^k) + 4(2^k)$$
$$= 9(5p - 2^k) + 4(2^k) \quad \text{using assumption } [\star]$$
$$= 45p - 9(2^k) + 4(2^k)$$
$$= 45p - 5(2^k)$$
$$= 5(9p - 2^k)$$
$$= 5q \text{ which is a multiple of 5.}$$

$\therefore P(k + 2)$ is true.

So $P(n)$ is true by mathematical induction.

> **Hint**
> Proving $P(k + 2)$ is true means truth of $P(k) \Rightarrow$ truth of $P(k + 2)$. As $P(1)$ is true, then the statement is true for all odd numbers n.

Sigma notation

$$\sum_{r=1}^{n} T_r \text{ means } T_1 + T_2 + T_3 + T_4 + \cdots + T_{n-1} + T_n$$

For example:

$$\sum_{k=3}^{6} \frac{1}{3^{k-2}} = \frac{1}{3^{3-2}} + \frac{1}{3^{4-2}} + \frac{1}{3^{5-2}} + \frac{1}{3^{6-2}}$$

$$= \frac{1}{3^1} + \frac{1}{3^2} + \frac{1}{3^3} + \frac{1}{3^4}$$

$$= \frac{40}{81}$$

Example 9

Prove by mathematical induction

$$\sum_{k=1}^{N} \frac{1}{(2k+1)(2k-1)} = \frac{N}{2N+1},$$

where $k \in \mathbb{N}$.

Solution

Write out the **series** to see the pattern and where the series stops.

$\displaystyle\sum_{k=1}^{N} \frac{1}{(2k+1)(2k-1)}$ means

$$\frac{1}{[2(1)+1][2(1)-1]} + \frac{1}{[2(2)+1][2(2)-1]} + \frac{1}{[2(3)+1][2(3)-1]} + \cdots + \frac{1}{[2N+1][2N-1]}$$

$$= \frac{1}{3 \times 1} + \frac{1}{5 \times 3} + \frac{1}{7 \times 5} + \cdots + \frac{1}{(2N+1)(2N-1)}.$$

9780170459266

TOPIC SUMMARY

Proof:

Let $P(N)$ be the proposition that $\dfrac{1}{3 \times 1} + \dfrac{1}{5 \times 3} + \dfrac{1}{7 \times 5} + \cdots + \dfrac{1}{(2N+1)(2N-1)} = \dfrac{N}{2N+1}, \forall N \in \mathbb{N}$

Prove true for $P(1)$:

LHS $= \dfrac{1}{3 \times 1}$ RHS $= \dfrac{1}{2(1)+1}$

     $= \dfrac{1}{3}$ $= \dfrac{1}{3}$

LHS = RHS so $P(1)$ is true.

Assume $P(k)$ is true for some $k \in \mathbb{N}$.

$$\dfrac{1}{3 \times 1} + \dfrac{1}{5 \times 3} + \dfrac{1}{7 \times 5} + \cdots + \dfrac{1}{(2k+1)(2k-1)} = \dfrac{k}{2k+1} \quad [*]$$

RTP: $P(k+1)$ is true,

that is, $\dfrac{1}{3 \times 1} + \dfrac{1}{5 \times 3} + \dfrac{1}{7 \times 5} + \cdots + \dfrac{1}{(2k+1)(2k-1)} + \dfrac{1}{(2(k+1)+1)(2(k+1)-1)} = \dfrac{(k+1)}{2(k+1)+1}$

which simplifies to:

$$\dfrac{1}{3 \times 1} + \dfrac{1}{5 \times 3} + \dfrac{1}{7 \times 5} + \cdots + \dfrac{1}{(2k+1)(2k-1)} + \dfrac{1}{(2k+3)(2k+1)} = \dfrac{k+1}{2k+3}.$$

Proof:

LHS of $P(k+1) = \dfrac{1}{3 \times 1} + \dfrac{1}{5 \times 3} + \dfrac{1}{7 \times 5} + \cdots + \dfrac{1}{(2k+1)(2k-1)} + \dfrac{1}{(2k+3)(2k+1)}$

$= \dfrac{k}{2k+1} + \dfrac{1}{(2k+3)(2k+1)}$ using $P(k)$ $[*]$

$= \dfrac{k(2k+3)+1}{(2k+3)(2k+1)}$

$= \dfrac{2k^2 + 3k + 1}{(2k+3)(2k+1)}$

$= \dfrac{(k+1)(2k+1)}{(2k+3)(2k+1)}$

$= \dfrac{k+1}{2k+3}$

$=$ RHS of $P(k+1)$

So $P(n)$ is true by mathematical induction.

Inequality proofs by induction

Example 10

Prove by mathematical induction:

$$2^n > n^2,$$

where $n \in \mathbb{N}$, $n \geq 5$.

Solution

Let $P(n)$ be the proposition that $2^n > n^2$, where $n \in \mathbb{N}$, $n \geq 5$.

Prove true for $P(5)$:

LHS $= 2^5$ RHS $= 5^2$

     $= 32$ $= 25$

Since LHS > RHS, then $P(5)$ is true.

Assume $P(k)$ true for some $k \in \mathbb{N}$, $k \geq 5$,

$2^k > k^2$ [*]

RTP: $P(k + 1)$ is true, $2^{k+1} > (k + 1)^2$

Proof:

LHS of $P(k + 1) = 2^{k+1}$

                $= 2^k \times 2^1$

                $> k^2 \times 2$ using $P(k)$ [*]

                $= k^2 + k^2$

Now $k^2 > 2k + 1$ since $k \geq 5$, so

LHS of $P(k + 1) > k^2 + (2k + 1)$

LHS of $P(k + 1) > (k + 1)^2$, which is RHS of $P(k + 1)$.

So $P(n)$ is true by mathematical induction.

Calculus proof by induction

Example 11

Prove by mathematical induction

$$\frac{d}{dx}(x^n) = nx^{n-1}, n \in \mathbb{N}, n \geq 1.$$

Solution

Let $P(n)$ be the proposition that

$\frac{d}{dx}(x^n) = nx^{n-1}, n \in \mathbb{N}$.

Prove true for $P(1)$:

LHS $= \frac{d}{dx}(x^1)$

     $= 1$

The gradient of $y = x$ is 1.

RHS $= 1 \times x^{1-1}$

     $= 1 \times x^0$

     $= 1$

LHS = RHS so $P(1)$ is true.

Assume $P(k)$ true for some $k \in \mathbb{N}$.

$\frac{d}{dx}(x^k) = kx^{k-1}$ [*]

RTP $P(k + 1)$ is true: $\frac{d}{dx}(x^{k+1}) = (k + 1)x^k$

Proof:

LHS of $P(k + 1) = \frac{d}{dx}(x^{k+1})$

                $= \frac{d}{dx}(x^k x)$

                $= x^k \frac{d}{dx}(x) + x \frac{d}{dx}(x^k)$ (product rule)

                $= x^k \times 1 + x \times kx^{k-1}$ using $P(k)$ [*]

                $= x^k + kx^k$

                $= x^k(1 + k)$

                $= (k + 1)x^k$, which is RHS of $P(k + 1)$.

So $P(n)$ is true by mathematical induction.

Recursive formula proof by induction

Example 12

Given $T_1 = 1$, $T_2 = 5$ and $T_n = 5T_{n-1} - 6T_{n-2}$, prove that

$$T_n = 3^n - 2^n$$

is true $\forall n \in \mathbb{N}$, $n \geq 3$ by mathematical induction.

Solution

Let $P(n)$ be the proposition that if $T_1 = 1$, $T_2 = 5$ and $T_n = 5T_{n-1} - 6T_{n-2}$,

then $T_n = 3^n - 2^n$ is true $\forall n \in \mathbb{N}$, $n \geq 3$.

Prove true for $P(3)$,

given $T_1 = 1$, $T_2 = 5$, $T_3 = 5 \times 5 - 6 \times 1 = 19$.

Using the formula $T_n = 3^n - 2^n$,

$T_3 = 3^3 - 2^3 = 19$, which is consistent.

$\therefore$ $P(3)$ is true.

Assume $P(k)$ and $P(k-1)$ true for some $k \in \mathbb{N}$, $k \geq 3$,

that is, given $T_k = 5T_{k-1} - 6T_{k-2}$, then $T_k = 3^k - 2^k$ and $T_{k-1} = 3^{k-1} - 2^{k-1}$.   [*]

RTP $P(k+1)$ is true:

that is, given $T_{k+1} = 5T_k - 6T_{k-1}$, then $T_{k+1} = 3^{k+1} - 2^{k+1}$.

Proof:

$T_{k+1} = 5T_k - 6T_{k-1}$

Now use both $T_k = 3^k - 2^k$ and $T_{k-1} = 3^{k-1} - 2^{k-1}$ [*] to substitute.

$$
\begin{aligned}
T_{k+1} &= 5(3^k - 2^k) - 6(3^{k-1} - 2^{k-1}) \\
&= 5(3^k) - 5(2^k) - 6(3^{k-1}) + 6(2^{k-1}) \\
&= 5(3^k) - 5(2^k) - 2 \times 3 \times 3^{k-1} + 3 \times 2 \times 2^{k-1} \\
&= 5(3^k) - 5(2^k) - 2(3^k) + 3(2^k) \\
&= 3(3^k) - 2(2^k) \\
&= 3^{k+1} - 2^{k+1}, \text{ which is the required expression.}
\end{aligned}
$$

So $P(n)$ is true by mathematical induction.

Practice set 1

Multiple-choice questions

Solutions start on page 23.

Question 1 ⬤⬛⬛

Which of the following statements is always true for $n \in \mathbb{N}$?

A $n(n + 1)(n - 1)$ is divisible by 8

B $(n + 1)^2 - n^2$ is even

C $n(n + 1)(n + 2)$ is divisible by 3

D $2^n - 1$ is always prime

Question 2 ⬤⬛⬛

Find the statement that is contrapositive to:

> 'If tigers are not carnivorous, then they do not eat humans.'

A If tigers are carnivorous, then they eat humans.

B If tigers eat humans, then they are carnivorous.

C If tigers do not eat humans, then they are not carnivorous.

D Tigers eat humans if and only if they are carnivorous.

Question 3 ⬤⬤⬛

$P(n)$ is the proposition that for $n \in \mathbb{N}$, $x^n - y^n$ is divisible by $x^2 - y^2$, where n is even.

Find the statement for $P(k + 2)$.

A $x^{k+2} - y^{k+2}$ is divisible by $x^4 - y^4$

B $(x + 2)^k - (y + 2)^k$ is divisible by $x^2 - y^2$

C $(x + 2)^k - (y + 2)^k$ is divisible by $(x + 2)^2 - (y + 2)^2$

D $x^{k+2} - y^{k+2}$ is divisible by $x^2 - y^2$

Question 4 ⬤⬤⬛

Find the converse to the statement:

> 'If you wear sunscreen, then you do not get sunburn.'

A If you do not get sunburn, then you wear sunscreen.

B If you do not wear sunscreen, then you get sunburn.

C If you get sunburn, then you do not wear sunscreen.

D If you get sunburn, then you wear sunscreen.

Question 5 ⬤⬛⬛

Find the counterexample to the statement:

> 'All boys are good at Mathematics.'

A All boys are not good at Mathematics.

B Some girls are good at Mathematics.

C If you are good at Mathematics, then you are a boy.

D James, who is a boy, is not good at Mathematics.

Question 6 ⬤◻◻

What is the contrapositive of $\neg R \Rightarrow K$?

A $K \Rightarrow \neg R$ **B** $\neg K \Rightarrow \neg R$ **C** $\neg K \Rightarrow R$ **D** $\neg K \Leftrightarrow R$

Question 7 ⬤⬤◻

Find the correct mathematical notation for the statement:

'For all integers x, there exist rational numbers y and w such that x is the product of y and w.'

A $\forall x \in \mathbb{N}, \exists y, w \in \mathbb{R} : x = yw.$ **B** $\forall x \in \mathbb{Q}, \exists y, w \in \mathbb{R} : x = \dfrac{y}{w}.$

C $\forall x \in \mathbb{Z}, \exists y, w \in \mathbb{Q} : x = yw.$ **D** $\forall x \in \mathbb{N}, \exists y, w \in \mathbb{Q} : x = \dfrac{y}{w}.$

Question 8 ⬤⬤◻

Which of the following statements is always true?

A If the square of a natural number is even, then the number is even.

B $u > \dfrac{1}{u}$

C $x^n + 1$ is divisible by $x + 1$ for $n \in \mathbb{N}$.

D If $a < b$, then $a^2 < b^2$.

Question 9 ⬤⬤◻

Consider the following two statements:

P: A number is divisible by 2.

Q: A number is composite.

Which of the following statements is true?

A $P \Rightarrow Q$

B $Q \Rightarrow P$

C $\neg Q \Rightarrow \neg P$

D None of the above

Question 10 ⬤◻◻

Consider the following two statements.

X: If politicians are not honest, then they are criticised.

Y: If politicians are not criticised, then they are honest.

Which of the following is correct?

A Y is the converse of X **B** Y is the contrapositive of X

C Y is the negation of X **D** Y is a counterexample of X

Question 11 ⬤◻◻

If $A \Rightarrow \neg B$ is false, which of the following is always false?

A $\neg A \Rightarrow B$ **B** $\neg B \Rightarrow A$ **C** $B \Rightarrow \neg A$ **D** $\neg A \Rightarrow \neg B$

Question 12

Which of the following does not have a counterexample?

A $\forall x, y \in \mathbb{R}, \exists n \in \mathbb{R}$ such that $x^2 + y^2 = n^2$

B $\forall x, y \in \mathbb{Z}, \exists z \in \mathbb{Q}$ such that $x = zy$

C $\forall x, y \in \mathbb{Q}, \exists q \in \mathbb{Z}$ such that $x^q = y$

D $\forall x, y \in \mathbb{N}, \exists k \in \mathbb{N}$ such that $x - y = k$

Question 13

Consider the statement:

'If n is divisible by 2 and 3, then n is divisible by 6.'

To start proving this by contradiction, what assumption is made?

A Assume n is divisible by 2 and 3 but not divisible by 6.

B Assume n is divisible by 2 but not divisible by 3.

C Assume n is divisible by 3 but not divisible by 2.

D Assume n is not divisible by 2 and 3 but is divisible by 6.

Question 14

Which type of proof should be used to prove that $\forall n \in \mathbb{N}, 3^n > n^3$, where $n > 3$?

A Proof by contrapositive

B Proof by contradiction

C Proof by induction

D Proof by counterexample

Question 15

Which of the following inequalities is always true if $a > b$?

A $\dfrac{1}{a} < \dfrac{1}{b}$

B $a^3 > b^3$

C $\sqrt{a} > \sqrt{b}$

D $\dfrac{1}{a + b} > 0$

Question 16

Find the negation of the statement:

'All even square numbers are divisible by 4.'

A All square numbers divisible by 4 are even.

B All square numbers not divisible by 4 are even.

C There exists an even square number that is not divisible by 4.

D There does not exist an even square number that is divisible by 4.

Question 17

Which type of proof is commonly used to disprove that a statement is always true?

A Proof by converse

B Proof by contradiction

C Proof by counterexample

D Proof by negation

Question 18 ◐◐◑

The following statement is not true:

'There is at least one girl who plays tennis.'

Which statement is true then?

A All girls play tennis.

B There is at least one girl who does not play tennis.

C All students who play tennis are girls.

D All girls do not play tennis.

Question 19 ©NESA 2020 HSC EXAM, QUESTION 7 ●●●

Consider the proposition:

'If $2^n - 1$ is not prime, then n is not prime'.

Given that each of the following statements is true, which statement disproves the proposition?

A $2^5 - 1$ is prime

B $2^6 - 1$ is divisible by 9

C $2^7 - 1$ is prime

D $2^{11} - 1$ is divisible by 23

Question 20 ©NESA 2020 HSC EXAM, QUESTION 8 ●●●

Consider the statement:

'If n is even, then if n is a multiple of 3, then n is a multiple of 6'.

Which of the following is the negation of this statement?

A n is odd and n is not a multiple of 3 or 6.

B n is even and n is a multiple of 3 but not a multiple of 6.

C If n is even, then n is not a multiple of 3 and n is not a multiple of 6.

D If n is odd, then if n is not a multiple of 3 then n is not a multiple of 6.

Practice set 2

Short-answer questions

Solutions start on page 25.

Question 1 (2 marks)

Consider this statement:

'If I eat food, then I brush my teeth.'

a State the converse of the statement.   1 mark

b State the contrapositive of the statement.   1 mark

Question 2 (5 marks)

Consider this statement:

'All integers are rational.'

It can also be written this way:

'If x is an integer, then it is rational.'

a Write down the negation of this statement.   1 mark

b The statement can be written in mathematical notation as $\forall x \in \mathbb{Z}, x \in \mathbb{Q}$.   2 marks

Write the negation of the statement in mathematical notation.

c Write the negation of this statement in mathematical notation: 2 marks

$$\forall x \in \mathbb{N} : \frac{1}{x} \leq 0.$$

Question 3 (2 marks)

a Rewrite this statement in words: 1 mark

$$\forall x, y, n \in \mathbb{N}, \exists z \in \mathbb{N} : z^n = x^n + y^n.$$

b Prove that this statement is not true by giving a counterexample.   1 mark

Question 4 (6 marks)

Consider these two statements:

P: The quadrilateral has 4 equal angles.

Q: The quadrilateral is a square.

$P \Rightarrow Q$ means 'If the quadrilateral has 4 equal angles, then it is a square.'

a Write $Q \Rightarrow P$ in words.   2 marks

b Write $P \Leftrightarrow Q$ in words and state whether it is true, giving reasons.   2 marks

c Write $\neg Q \Leftrightarrow \neg P$ in words and state whether it is true, giving reasons.   2 marks

Question 5 (5 marks)

Prove each statement is true for natural numbers n.

a The sum of 3 consecutive numbers is divisible by 3.   2 marks

b If n is even, then $n^2 - 4n + 3$ is odd.   3 marks

Question 6 (3 marks)

Prove that every odd number greater than 1 is a difference of 2 consecutive square numbers.   3 marks

Question 7 (3 marks)

Prove by the contrapositive the statement: 3 marks

$$\forall\, n \in \mathbb{N}, \text{ if } 5n + 11 \text{ is even, then } n \text{ is odd.}$$

Question 8 (4 marks)

Consider a Pythagorean triad (a, b, c) where $a < b < c$ such that

$$a = pq, \; b = \frac{p^2 - q^2}{2}, \; c = \frac{p^2 + q^2}{2},$$

where $p > q$ and p, q are odd.

Prove that $\dfrac{(c - a)(c - b)}{2}$ is always a square number.   4 marks

Question 9 (2 marks)

Determine the number of solutions to the inequality $x \in \mathbb{Z} : x^2 < 10$.   2 marks

Question 10 (3 marks)

Prove that $\forall\, n \in \mathbb{N}, n^2 + 2$ is not divisible by 4.   3 marks

Question 11 (7 marks)

Prove each formula for $n, k \in \mathbb{Z}$ where $n \geq k \geq 0$.

a $\dbinom{n}{k} = \dbinom{n}{n - k}$ 2 marks

b $\displaystyle\sum_{k=1}^{n} k = \dbinom{n + 1}{2}$ 2 marks

c $\dbinom{2n}{2} = 2\dbinom{n}{2} + n^2$ 3 marks

Question 12 (4 marks)

a Expand $(1 + x)^n$ in ascending powers of x.   1 mark

b By differentiating the result above and by making a suitable substitution, hence prove   3 marks

$$\sum_{k=1}^{n} k \binom{n}{k} = n2^{n-1}.$$

Question 13 (4 marks)

Prove each inequality.

a Prove that $\dfrac{m}{n} + \dfrac{n}{m} \geq 2$ for $m, n > 0$.   2 marks

b Hence prove that $p^4 + q^4 + r^4 \geq p^2q^2 + q^2r^2 + r^2p^2$.   2 marks

Question 14 (3 marks) ●●○

Prove that it is impossible to find 4 distinct real numbers a, b, c and d such that

3 marks

$$ab + cd = ad + bc.$$

Question 15 (6 marks) ●●○

Prove each statement by induction.

a $\displaystyle\sum_{r=1}^{n}\frac{1}{r(r+1)} = \frac{n}{n+1}, n \in \mathbb{N}$.

3 marks

b $7^n + 11^n$ is divisible by 9 if n is odd.

3 marks

Question 16 (6 marks) ●●●

Prove each statement by induction.

a Given a function f such that $f(x + y) = f(x) + f(y)$, $\forall n \in \mathbb{N}, f(nx) = nf(x)$.

3 marks

b $\forall n \in \mathbb{N}, (x + 1)^n - nx - 1$ is divisible by x^2.

3 marks

Question 17 (7 marks) ●●●

a Prove that $\sin(A + B) - \sin(A - B) = 2\cos A \sin B$.

3 marks

b Prove by induction that $\forall n \in \mathbb{N}$,

4 marks

$$\cos\alpha + \cos 3\alpha + \cos 5\alpha + \cdots + \cos(2n - 1)\alpha = \frac{\sin 2n\alpha}{2\sin\alpha}.$$

Question 18 (3 marks) ●●○

Given that $T_1 = 5$ and $T_n = 1 + 2T_{n-1}$ for $n \geq 2$, prove that $T_n = 6 \times 2^{n-1} - 1$ for all positive integers n.

3 marks

Question 19 (7 marks) ●●●

The number of people at a party is n. Each person shakes hands once with everyone else in the room.

a Complete this table.

2 marks

Number of people, n	1	2	3	4	5
Number of handshakes, $h(n)$	0	1	3		

b Verify when $n = 6$ that the formula for the number of handshakes is $h(n) = \dfrac{n(n-1)}{2}$.

2 marks

c Prove $h(n) = \dfrac{n(n-1)}{2}$ by induction.

3 marks

[Hint: consider an extra person joining the room.]

Question 20 (5 marks) ⬤⬤⬤

Consider the graph of $y = \dfrac{1}{x+1}$, the secant through the points on the graph where $x = 0$ and $x = n$, and the line $y = \dfrac{1}{n+1}$.

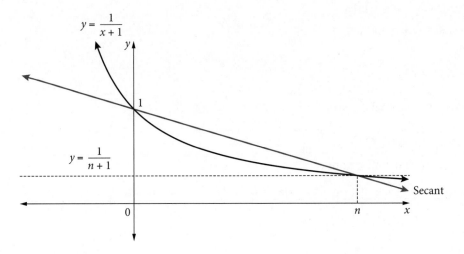

By considering areas, prove that   5 marks

$$e^n < (n+1)^{n+1} < e^{\frac{n(n+2)}{2}}.$$

Question 21 (6 marks) ©NESA 2012 HSC EXAM, QUESTION 15(a) ⬤⬤⬤

a   Prove that   1 mark

$$\sqrt{ab} \le \frac{a+b}{2},$$

where $a \ge 0$ and $b \ge 0$.

b   If $1 \le x \le y$, show that $x(y - x + 1) \ge y$.   2 marks

c   Let n and j be positive integers with $1 \le j \le n$.   2 marks

Prove that

$$\sqrt{n} \le \sqrt{j(n - j + 1)} \le \frac{n+1}{2}.$$

d   For integers $n \ge 1$, prove that   1 mark

$$\left(\sqrt{n}\right)^n \le n! \le \left(\frac{n+1}{2}\right)^n.$$

Practice set 1

Worked solutions

1 C

Every third number in the set of natural numbers is a multiple of 3, so for any 3 consecutive numbers, one of them must be a multiple of 3. Therefore, the product of any 3 consecutive numbers must be divisible by 3. Therefore, $n(n + 1)(n + 2)$ is divisible by 3. Counterexamples can be found for the other options: A ($n = 2$), B ($n = 5$) and D ($n = 4$).

2 B

To find the contrapositive to an if-then statement, swap the 'if' and 'then' phrases and negate each.

So 'If tigers are not carnivorous, then they do not eat humans' becomes:

'If tigers eat humans, then they are carnivorous.'

3 D

Replace n with $(k + 2)$ so $x^{k+2} - y^{k+2}$ is divisible by $x^2 - y^2$.

4 A

The converse of $P \Rightarrow Q$ is $Q \Rightarrow P$.

Swap the if-then phrases.

Therefore the converse is:

'If you do not get sunburn, then you wear sunscreen'.

5 D

Choose a boy who is not good at Mathematics. James is that boy. Note: Statement A contradicts the claim that "all boys are good at Mathematics", but it is not a counterexample since it doesn't identify a specific boy.

6 C

The contrapositive of $P \Rightarrow Q$ is $\neg Q \Rightarrow \neg P$ so the contrapositive of $\neg R \Rightarrow K$ is $\neg K \Rightarrow R$.

7 C

In words	In mathematical notation
For all integers x,	$\forall x \in \mathbb{Z}$
there exist rational numbers y and w	$\exists y, w \in \mathbb{Q}$
such that x is the product of y and w.	$: x = yw$

8 A

The square of an even number is always even.

The square of an odd number is always odd.

If the square of a natural number is even, then the number must be even.

Counterexamples can be found for options: B $\left(u = \frac{1}{5}\right)$, C ($n = 2$) and D ($a = -1, b = 0$).

9 D

None of the options A, B or C are correct as the number 2 is prime AND divisible by 2.

10 B

Let 'If politicians are not honest, then they are criticised' be $\neg P \Rightarrow Q$.

Then 'If politicians are not criticised, then they are honest' is $\neg Q \Rightarrow P$, that is, the contrapositive.

11 C

A statement and its contrapositive are equivalent, so if one is false, the other is also false. $(A \Rightarrow \neg B) \Leftrightarrow (B \Rightarrow \neg A)$.

12 A

Given x, y are real, there exists a value of n that is real such that $x^2 + y^2 = n^2$.

Since n^2 is non-negative, it is possible to find a real solution for n.

All other options have counterexamples:

B: $x = 5$, $y = 0$, z is undefined

C: $x = 25$, $y = 5$, which would require $q = \dfrac{1}{2}$ (not an integer)

D: $x = 3$, $y = 10$, which would require $k = -7$

(not a natural number)

13 A

Assume the negation is true.

We are trying to prove that n is divisible by 6 so assume n is not divisible by 6.

14 C

Because we must prove the statement is true for all integer values of $n > 3$, induction is the most straightforward approach. It is possible to prove this statement by other methods, but much more difficult.

15 B

Some of these inequalities are true only if a and b are positive. Consider if a and/or b are negative. If always true then that must include negative numbers and 0.

By considering all cases for the signs of a and b, it can be shown that $a^3 > b^3$ is true. Counterexamples can be found for the other options:

For option A, consider $a = 1$, $b = -1$, $a > b$. For options C and D, consider $a = -2$, $b = -3$, $a > b$.

A $\dfrac{1}{a} < \dfrac{1}{b}$ $1 < -1$ False

C $\sqrt{a} > \sqrt{b}$ $\sqrt{-2} > \sqrt{-3}$ False, undefined

D $\dfrac{1}{a+b} > 0$ $-\dfrac{1}{5} > 0$ False

16 C

The negation of "All are" is "at least one is not" or "some are not".

17 C

Only one case is required to prove a statement of this form is not true.

18 D

As in probability, the negation of "at least one" is "none".

'No girls play tennis', or 'All girls do not play tennis'.

It might appear that statement B is also true, but consider the case where there are no girls at all.

19 D

$2^{11} - 1$ is not prime as it is divisible by 23, but $n = 11$ is prime.

20 B

Add 'not' to negate a statement:

'If n is even, then if n is a multiple of 3, then n is *not* a multiple of 6'.

9780170459266

Practice set 2

Worked solutions

Question 1

a The converse: If I brush my teeth, then I eat food.

b The contrapositive: If I don't brush my teeth, then I don't eat food.

Question 2

a The negation:

'Not all integers are rational'
or 'At least one integer is not irrational'
or 'Some integers are not irrational.'
(This statement is false, by the way.)

b The negation: 'There is at least one integer that is not rational'

$\exists x \in \mathbb{Z} : x \notin \mathbb{Q}$

(This statement is false, by the way.)

c The negation: All natural numbers have positive reciprocals.

$\exists x \in \mathbb{N} : \dfrac{1}{x} > 0$

Question 3

a For all natural numbers x, y and n, there exists a natural number z such that $z^n = x^n + y^n$.

b Counterexample: Let $x = 1$, $y = 2$, $n = 2$:

$z^2 = x^2 + y^2$
$= 1^2 + 2^2$
$= 5$
$z = \sqrt{5} \notin \mathbb{N}$

So the statement is false.

Question 4

a $Q \Rightarrow P$: If the quadrilateral is a square, then it has 4 equal angles.

b $P \Leftrightarrow Q$: The quadrilateral has 4 equal angles if and only if it is a square.

This is not true because a *rectangle* also has 4 equal angles, not only squares.

False since $P \Rightarrow Q$ is false.

c $\neg Q \Leftrightarrow \neg P$: If a quadrilateral is not a square, then it does not have 4 equal angles.

This is not true. The quadrilateral could be a rectangle (not a square), which has 4 equal angles.

The contrapositive of $P \Rightarrow Q$ is not true because the original $P \Rightarrow Q$ is not true.

Question 5

a RTP: The sum of 3 consecutive numbers is divisible by 3.

Proof: Let the numbers be n, $n + 1$, $n + 2$, where $n \in \mathbb{N}$.

Then their sum is:

$$n + (n + 1) + (n + 2) = 3n + 3$$
$$= 3(n + 1),$$

which is divisible by 3.

b RTP: If n is even, then $n^2 - 4n + 3$ is odd.

Proof: Let $n = 2M$ where M is an integer.

$$n^2 - 4n + 3 = (2M)^2 - 4(2M) + 3$$
$$= 4M^2 - 8M + 3$$
$$= 4M^2 - 8M + 2 + 1$$
$$= 2(2M^2 - 4M + 1) + 1,$$

which is odd because $2(2M^2 - 4M + 1)$ is even.

Question 6

RTP: Every odd number greater than 1 is a difference of 2 consecutive square numbers.

Consider the difference of 2 consecutive square numbers $(X + 1)^2 - X^2$ where $X \in \mathbb{N}$.

$$(X + 1)^2 - X^2 = X^2 + 2X + 1 - X^2$$
$$= 2X + 1, \text{ which is odd.}$$

So every odd number of the form $2X + 1$ can be written as a difference of 2 consecutive square numbers $(X + 1)^2 - X^2$.

But as $X \in \mathbb{N}$, then the first odd number this works for is when $X = 1$, $2X + 1 = 2 + 1 = 3$. So every odd number greater than 1 is a difference of 2 consecutive square numbers.

Note: This property can also be proved by induction.

Question 7

RTP: $\forall\, n \in \mathbb{N}$, if $5n + 11$ is even, then n is odd.

Proof by the contrapositive: If n is even, then $5n + 11$ is odd.

Proof: Let $n = 2X$ where $X \in \mathbb{N}$.

$$
\begin{aligned}
5n + 11 &= 5(2X) + 11 \\
&= 10X + 11 \\
&= 10X + 10 + 1 \\
&= 2(5X + 5) + 1,
\end{aligned}
$$

which is odd, as $2(5X + 5)$ is even.

Therefore the contrapositive is true, so the original statement is true:

$\forall\, n \in \mathbb{N}$, if $5n + 11$ is even, then n is odd.

Question 8

Pythagorean triad (a, b, c), where $a < b < c$ such that

$$a = pq,\ b = \frac{p^2 - q^2}{2},\ c = \frac{p^2 + q^2}{2},$$

where $p > q$ and p, q are odd.

RTP: $\dfrac{(c - a)(c - b)}{2}$ is always a perfect square.

Proof:

$$
\begin{aligned}
\frac{(c - a)(c - b)}{2} &= \frac{\left(\dfrac{p^2 + q^2}{2} - pq\right)\left(\dfrac{p^2 + q^2}{2} - \dfrac{p^2 - q^2}{2}\right)}{2} \\[2mm]
&= \frac{\left(\dfrac{p^2 - 2pq + q^2}{2}\right)\left(\dfrac{2q^2}{2}\right)}{2} \\[2mm]
&= \frac{\dfrac{(p - q)^2 q^2}{2}}{2} \\[2mm]
&= \frac{(p - q)^2 q^2}{4} \\[2mm]
&= \left(\frac{(p - q)q}{2}\right)^2
\end{aligned}
$$

To prove that this is a square number, we need to show that $(p - q)q$ is even so that it can be divided by the 2 in the denominator.

Since p, q are odd, then $p - q$ is even, so $(p - q)q$ is even, so

$$
\begin{aligned}
\frac{(c - a)(c - b)}{2} &= \left(\frac{(p - q)q}{2}\right)^2 \\[2mm]
&= \left(\frac{2X}{2}\right)^2 \qquad \text{for some } X \in \mathbb{N} \\[2mm]
&= X^2,
\end{aligned}
$$

which is a square number.   QED.

Question 9

$x \in \mathbb{Z} : x^2 < 10$ means

$-\sqrt{10} < x < \sqrt{10}$ for integer x

$-3.16\ldots < x < 3.16\ldots$

$\therefore x = -3, -2, -1, 0, 1, 2, 3$

Therefore, there are 7 solutions.

Question 10

RTP: $\forall\, n \in \mathbb{N}$, $n^2 + 2$ is not divisible by 4.

Consider 2 cases, whether n is even or odd:

Let n be even: then $n = 2P,\ P \in \mathbb{N}$ so

$$
\begin{aligned}
n^2 + 2 &= (2P)^2 + 2 \\
&= 4P^2 + 2, \text{ which is not divisible by 4} \\
&\qquad (4P^2 \text{ is divisible by 4, but 2 isn't)}.
\end{aligned}
$$

Let n be odd: then $n = 2P - 1,\ P \in \mathbb{N}$ so

$$
\begin{aligned}
n^2 + 2 &= (2P - 1)^2 + 2 \\
&= 4P^2 - 4P + 1 + 2 \\
&= 4P^2 - 4P + 3 \\
&= 4(P^2 - 1) + 3, \text{ which is not divisible by 4} \\
&\qquad (4(P^2 - 1) \text{ is divisible by 4,} \\
&\qquad \text{but 3 isn't)}.
\end{aligned}
$$

Therefore, $\forall\, n \in \mathbb{N}$, $n^2 + 2$ is not divisible by 4.

Question 11

a $\dbinom{n}{k} = \dbinom{n}{n - k}$

Proof:

$$
\begin{aligned}
\binom{n}{k} &= \frac{n!}{k!(n - k)!} \\[2mm]
\binom{n}{n - k} &= \frac{n!}{(n - k)!(n - (n - k))!} \\[2mm]
&= \frac{n!}{(n - k)!k!} \\[2mm]
&= \frac{n!}{k!(n - k)!} \\[2mm]
&= \binom{n}{k}, \text{ as required.}
\end{aligned}
$$

b $\displaystyle\sum_{k=1}^{n} k = \binom{n+1}{2}$

Proof:

$$\sum_{k=1}^{n} k = 1 + 2 + 3 + 4 + \cdots + n$$

$$= \frac{n}{2}(a + l) \qquad \text{Arithmetic series where}$$
$$\qquad\qquad\qquad a = 1, l = n, n \text{ terms}$$

$$= \frac{n}{2}(1 + n)$$

$$\binom{n+1}{2} = \frac{(n+1)!}{2!(n+1-2)!}$$

$$= \frac{(n+1)!}{2!(n-1)!}$$

$$= \frac{(n+1)n(n-1)!}{2!(n-1)!}$$

$$= \frac{n(n+1)}{2}$$

$$= \sum_{k=1}^{n} k, \text{ as required.}$$

c $\displaystyle\binom{2n}{2} = 2\binom{n}{2} + n^2$

Proof:

$$\binom{2n}{2} = \frac{(2n)!}{2!(2n-2)!}$$

$$= \frac{(2n)(2n-1)(2n-2)!}{2(2n-2)!}$$

$$= n(2n-1)$$

$$2\binom{n}{2} + n^2 = \frac{2 \times n!}{2!(n-2)!} + n^2$$

$$= \frac{2n(n-1)(n-2)!}{2(n-2)!} + n^2$$

$$= n(n-1) + n^2$$

$$= n^2 - n + n^2$$

$$= 2n^2 - n$$

$$= n(2n-1)$$

$$= \binom{2n}{2}, \text{ as required.}$$

Question 12

a $(1 + x)^n = {}^nC_0 x^0 + {}^nC_1 x^1 + {}^nC_2 x^2 + {}^nC_3 x^3 + \cdots + {}^nC_{n-1} x^{n-1} + {}^nC_n x^n$

b RTP:

$$\sum_{k=1}^{n} k\binom{n}{k} = n2^{n-1}, \text{ that is,}$$

$$1\binom{n}{1} + 2\binom{n}{2} + 3\binom{n}{3} + \cdots + (n-1)\binom{n}{n-1} + n\binom{n}{n} = n2^{n-1}.$$

Proof:

Differentiating both sides of the identity:

$$(1 + x)^n = {}^nC_0 x^0 + {}^nC_1 x^1 + {}^nC_2 x^2 + {}^nC_3 x^3 + \cdots + {}^nC_{n-1} x^{n-1} + {}^nC_n x^n$$

$$n(1 + x)^{n-1} = 0 + 1{}^nC_1 x^0 + 2{}^nC_2 x^1 + 3{}^nC_3 x^2 + \cdots + (n-1){}^nC_{n-1} x^{n-2} + n{}^nC_n x^{n-1}$$

Substituting $x = 1$:

$$n(1 + 1)^{n-1} = 1{}^nC_1(1)^0 + 2{}^nC_2(1)^1 + 3{}^nC_3(1)^2 + \cdots + (n-1){}^nC_{n-1}(1)^{n-2} + n{}^nC_n(1)^{n-1}$$

$$n(2)^{n-1} = 1{}^nC_1 + 2{}^nC_2 + 3{}^nC_3 + \cdots + (n-1){}^nC_{n-1} + n{}^nC_n$$

$$\therefore n \times 2^{n-1} = \binom{n}{1} + 2\binom{n}{2} + 3\binom{n}{3} + \cdots + n\binom{n}{n}$$

$$= \sum_{k=1}^{n} k\binom{n}{k}, \text{ as required.}$$

WORKED SOLUTIONS

Question 13

a RTP: $\dfrac{m}{n} + \dfrac{n}{m} \geq 2$ for $m, n > 0$.

Proof: Consider the difference:

$$\dfrac{m}{n} + \dfrac{n}{m} - 2 = \dfrac{m^2 + n^2 - 2mn}{mn}$$

$$= \dfrac{(m-n)^2}{mn}$$

$$\geq 0 \quad \text{since } (m-n)^2 \geq 0 \text{ and } mn > 0$$

Therefore $\dfrac{m}{n} + \dfrac{n}{m} \geq 2$.   QED.

b RTP: $p^4 + q^4 + r^4 \geq p^2q^2 + q^2r^2 + r^2p^2$

Proof: From part **a**:

$$\dfrac{m}{n} + \dfrac{n}{m} \geq 2$$

$$\dfrac{m^2 + n^2}{nm} \geq 2$$

$$m^2 + n^2 \geq 2mn$$

$$\therefore (p^2)^2 + (q^2)^2 \geq 2(p^2)(q^2) \qquad p^2 > 0, q^2 > 0$$

$$\therefore p^4 + q^4 \geq 2p^2q^2 \qquad [1]$$

Similarly,

$$q^4 + r^4 \geq 2q^2r^2 \qquad\qquad [2]$$

$$p^4 + r^4 \geq 2p^2r^2 \qquad\qquad [3]$$

Add all 3 inequalities together:

$$2p^4 + 2q^4 + 2r^4 \geq 2p^2q^2 + 2q^2r^2 + 2p^2r^2$$

$$p^4 + q^4 + r^4 \geq p^2q^2 + q^2r^2 + p^2r^2. \quad \text{QED.}$$

Question 14

RTP: $a, b, c, d \in \mathbb{R}$ do not exist such that $a \neq b \neq c \neq d$ and $ab + cd = ad + bc$.

Proof by contradiction: Assume that $a, b, c, d \in \mathbb{R}$ exist such that $a \neq b \neq c \neq d$ and $ab + cd = ad + bc$.

$$\text{Then } ab + cd = ad + bc$$

$$ab + cd - ad - bc = 0$$

$$ab - ad + cd - bc = 0$$

$$a(b - d) + c(d - b) = 0$$

$$a(b - d) - c(b - d) = 0$$

$$(b - d)(a - c) = 0$$

So $b = d$ or $a = c$.

Contradiction since $a \neq b \neq c \neq d$.

Therefore, $a, b, c, d \in \mathbb{R}$ do not exist such that $a \neq b \neq c \neq d$ and $ab + cd = ad + bc$.   QED.

Question 15

a Let $P(n)$ be the proposition that $\displaystyle\sum_{r=1}^{n}\frac{1}{r(r+1)} = \frac{n}{n+1}, n \in \mathbb{N}$,

that is, $\dfrac{1}{1 \times 2} + \dfrac{1}{2 \times 3} + \dfrac{1}{3 \times 4} + \cdots + \dfrac{1}{n(n+1)} = \dfrac{n}{n+1}, n \in \mathbb{N}$.

Prove $P(1)$ is true:

$\text{LHS} = \dfrac{1}{1 \times 2} = \dfrac{1}{2}$ $\text{RHS} = \dfrac{1}{1+1} = \dfrac{1}{2} = \text{LHS}$

So $P(1)$ is true.

Assume $P(k)$ is true for some $k \in \mathbb{N}$, that is,

$\dfrac{1}{1 \times 2} + \dfrac{1}{2 \times 3} + \dfrac{1}{3 \times 4} + \cdots + \dfrac{1}{k(k+1)} = \dfrac{k}{k+1}, k \in \mathbb{N}$ [*]

RTP: $P(k+1)$ is true, that is,

$\dfrac{1}{1 \times 2} + \dfrac{1}{2 \times 3} + \dfrac{1}{3 \times 4} + \cdots + \dfrac{1}{k(k+1)} + \dfrac{1}{(k+1)(k+2)} = \dfrac{k+1}{k+2}$.

Proof:

Consider the LHS of $P(k+1)$:

$$\dfrac{1}{1 \times 2} + \dfrac{1}{2 \times 3} + \dfrac{1}{3 \times 4} + \cdots + \dfrac{1}{k(k+1)} + \dfrac{1}{(k+1)(k+2)} = \dfrac{k}{k+1} + \dfrac{1}{(k+1)(k+2)} \quad \text{using } P(k) \text{ [*]}$$

$$= \dfrac{k(k+2)+1}{(k+1)(k+2)}$$

$$= \dfrac{k^2 + 2k + 1}{(k+1)(k+2)}$$

$$= \dfrac{(k+1)^2}{(k+1)(k+2)}$$

$$= \dfrac{k+1}{k+2}$$

$$= \text{RHS of } P(k+1)$$

So $P(k+1)$ is true.

Therefore $P(n)$ is true by mathematical induction.

b Let $P(n)$ be the proposition that $7^n + 11^n$ is divisible by 9 if n is odd.

Prove $P(1)$ true:

$7^1 + 11^1 = 7 + 11$
$\qquad = 18$

which is divisible by 9.

So $P(1)$ is true.

Assume $P(k)$ is true for some odd $k \in \mathbb{N}$, that is

$7^k + 11^k = 9X$ for some $X \in \mathbb{N}$
$\qquad 7^k = 9X - 11^k$ [*]

RTP: $P(k+2)$ is true: $7^{k+2} + 11^{k+2} = 9Y$ for some $Y \in \mathbb{N}$.

Proof:

Consider the LHS of $P(k+2)$:

$7^{k+2} + 11^{k+2} = 7^k 7^2 + 11^k 11^2$
$\qquad = 49(7^k) + 121(11^k)$
$\qquad = 49(9X - 11^k) + 121(11^k)$
$\qquad = 49(9X) - 49(11^k) + 121(11^k) \quad \text{using } P(k) \text{ [*]}$
$\qquad = 9(49X) + 72(11^k)$
$\qquad = 9(49X + 8(11^k))$
$\qquad = 9Y$, which is divisible by 9.

So $P(k+2)$ is true.

Therefore, $P(n)$ is true by mathematical induction.

Question 16

a Let $P(n)$ be the proposition that given a function f such that $f(x + y) = f(x) + f(y), f(nx) = nf(x)$.

Prove $P(1)$ is true:

LHS $= f(1x) = f(x)$

RHS $= 1f(x) = f(x)$

$\therefore$ LHS = RHS

So $P(1)$ is true.

Assume $P(k)$ is true for some $k \in \mathbb{N}$, that is given a function f such that $f(x + y) = f(x) + f(y)$,
$f(kx) = kf(x)$.   [*]

RTP: $P(k + 1)$ is true, that is, given a function f such that $f(x + y) = f(x) + f(y), f([k + 1]x) = (k + 1)f(x)$.

Proof:

Consider the LHS of $P(k + 1)$:

$$
\begin{aligned}
f([k + 1]x) &= f(kx + x) \\
&= f(kx) + f(x) \quad \text{using } f(x + y) = f(x) + f(y) \\
&= kf(x) + f(x) \quad \text{using } P(k) \; [*] \\
&= (k + 1)f(x) \\
&= \text{RHS}
\end{aligned}
$$

So $P(k + 1)$ is true.

Therefore, $P(n)$ is true by mathematical induction.

b Let $P(n)$ be the proposition that: $(x + 1)^n - nx - 1$ is divisible by x^2.

Prove $P(1)$ is true:

$$
\begin{aligned}
(x + 1)^1 - 1x - 1 &= x + 1 - x - 1 \\
&= 0, \text{ which is divisible by } x^2.
\end{aligned}
$$

So $P(1)$ is true.

Assume $P(k)$ is true for some $k \in \mathbb{N}$, that is,

for some $k \in \mathbb{N}$, $(x + 1)^k - kx - 1$ is divisible by x^2, that is,

$(x + 1)^k - kx - 1 = x^2 Q(x)$, where $Q(x)$ is some polynomial.

$$\therefore (x + 1)^k = x^2 Q(x) + kx + 1 \quad [*]$$

RTP: $P(k + 1)$ is true, that is, $(x + 1)^{k+1} - (k + 1)x - 1 = x^2 A(x)$, where $A(x)$ is some polynomial.

Proof:

Consider the LHS of $P(k + 1)$:

$$
\begin{aligned}
(x + 1)^{k+1} - (k + 1)x - 1 &= (x + 1)^k (x + 1)^1 - kx - x - 1 \\
&= [x^2 Q(x) + kx + 1](x + 1) - kx - x - 1 \quad \text{from } [*] \\
&= x^3 Q(x) + kx^2 + x + x^2 Q(x) + kx + 1 - kx - x - 1 \\
&= x^3 Q(x) + kx^2 + x^2 Q(x) \\
&= x^2 [xQ(x) + k + Q(x)] \\
&= x^2 A(x), \text{ where } A(x) \text{ is some polynomial.} \\
&= \text{RHS}
\end{aligned}
$$

So $P(k + 1)$ is true.

Therefore, $P(n)$ is true by mathematical induction.

Question 17

a Proof: Expanding:

$$\sin(A + B) - \sin(A - B) = \sin A \cos B + \cos A \sin B - (\sin A \cos B - \cos A \sin B)$$

$$= \cancel{\sin A \cos B} + \cos A \sin B - \cancel{\sin A \cos B} + \cos A \sin B$$

$$= 2 \cos A \sin B$$

b Let $P(n)$ be the proposition that

$$\cos \alpha + \cos 3\alpha + \cos 5\alpha + \cdots + \cos(2n - 1)\alpha = \frac{\sin 2n\alpha}{2 \sin \alpha}$$

Prove $P(1)$ is true:

LHS $= \cos \alpha$

RHS $= \dfrac{\sin 2\alpha}{2 \sin \alpha}$

$\quad\ = \dfrac{2\cancel{\sin \alpha} \cos \alpha}{2\cancel{\sin \alpha}}$

$\quad\ = \cos \alpha$

$\quad\ =$ LHS

$\therefore\ P(1)$ is true.

Assume $P(k)$ is true for some $k \in \mathbb{N}$, that is,

for some $k \in \mathbb{N}$,

$$\cos \alpha + \cos 3\alpha + \cos 5\alpha + \cdots + \cos(2k - 1)\alpha = \frac{\sin 2k\alpha}{2 \sin \alpha}. \quad [*]$$

RTP: $P(k + 1)$ is true, that is,

$$\cos \alpha + \cos 3\alpha + \cos 5\alpha + \cdots + \cos(2k - 1)\alpha + \cos(2[k + 1] - 1)\alpha = \frac{\sin 2(k + 1)\alpha}{2 \sin \alpha}$$

$$\cos \alpha + \cos 3\alpha + \cos 5\alpha + \cdots + \cos(2k - 1)\alpha + \cos(2k + 1)\alpha = \frac{\sin 2(k + 1)\alpha}{2 \sin \alpha}$$

Proof:

Consider the LHS of $P(k + 1)$:

$$\cos\alpha + \cos 3\alpha + \cos 5\alpha + \cdots + \cos(2k - 1)\alpha + \cos(2k + 1)\alpha$$

$$= \frac{\sin 2k\alpha}{2 \sin \alpha} + \cos(2k + 1)\alpha \quad \text{using } P(k)\ [*]$$

$$= \frac{\sin 2k\alpha + 2\cos(2k + 1)\alpha \sin \alpha}{2 \sin \alpha}$$

$$= \frac{\sin 2k\alpha + \sin[(2k + 1)\alpha + \alpha] - \sin[(2k + 1)\alpha - \alpha]}{2 \sin \alpha}$$

(using our formula from part **a**:
$2 \cos A \sin B = \sin(A + B) - \sin(A - B)$)

$$= \frac{\cancel{\sin 2k\alpha} + \sin[2k\alpha + 2\alpha] - \cancel{\sin 2k\alpha}}{2 \sin \alpha}$$

$$= \frac{\sin 2(k + 1)\alpha}{2 \sin \alpha}$$

$$=$$ RHS

So $P(k + 1)$ is true.

Therefore, $P(n)$ is true by mathematical induction.

Question 18

Given that $T_1 = 5$ and $T_n = 1 + 2T_{n-1}$ for $n \geq 2$, prove that $T_n = 6 \times 2^{n-1} - 1$ for all positive integers n.

Let $P(n)$ be the proposition that $T_n = 6 \times 2^{n-1} - 1$. Prove $P(1)$ is true:

Using the formula:

$T_1 = 6 \times 2^{1-1} - 1$
$\quad = 6 \times 2^0 - 1$
$\quad = 6 \times 1 - 1$
$\quad = 5$

$\therefore P(1)$ is true.

Assume $P(k)$ is true for some $k \in \mathbb{N}$, that is,

given that $T_1 = 5$ and $T_k = 1 + 2T_{k-1}$ for $k \geq 2$,

$T_k = 6 \times 2^{k-1} - 1.$ [*]

RTP: $P(k + 1)$ is true, that is,

given that $T_1 = 5$ and $T_{k+1} = 1 + 2T_k$,
$T_{k+1} = 6 \times 2^k - 1.$

Proof:

Consider $P(k + 1)$:

$T_{k+1} = 1 + 2T_k$
$\quad = 1 + 2(6 \times 2^{k-1} - 1)$ using $P(k)$ [*]
$\quad = 1 + 6(2^k) - 2$
$\quad = 6(2^k) - 1$
$\quad = $ RHS

So $P(k + 1)$ is true.

Therefore, $P(n)$ is true by mathematical induction.

Question 19

a

Number of people, n	1	2	3	4	5	6
Number of handshakes, $h(n)$	0	1	3	6	10	15

$\quad\quad\quad\quad\quad$ 1 $\quad$ 2 $\quad$ 3 $\quad$ 4 $\quad$ 5

b Using the pattern above, when $n = 6$, the number of handshakes should be 15.

$h(6) = \dfrac{6(6-1)}{2}$

$\quad = \dfrac{6(5)}{2}$

$\quad = 15$

This formula is correct for $n = 6$.

c Let $P(n)$ be the proposition that with n people there are $h(n) = \dfrac{n(n-1)}{2}$ handshakes.

Prove $P(1)$ is true:

If there is 1 person then there are 0 handshakes.

$h(1) = \dfrac{1(1-1)}{2}$

$\quad = \dfrac{1(0)}{2}$

$\quad = 0$

So $P(1)$ is true.

Assume $P(k)$ is true for some $k \in \mathbb{N}$, that is,

for some $k \in \mathbb{N}$, $h(k) = \dfrac{k(k-1)}{2}$.

RTP: $P(k + 1)$ is true, that is,

$h(k + 1) = \dfrac{(k+1)(k+1-1)}{2} = \dfrac{k(k+1)}{2}$

Proof:

Consider adding another person to the room, who shakes hands with k people so the number of handshakes for $(k + 1)$ people is

$h(k + 1) = \dfrac{k(k-1)}{2} + k$

$\quad = \dfrac{k(k-1) + 2k}{2}$

$\quad = \dfrac{k^2 - k + 2k}{2}$

$\quad = \dfrac{k^2 + k}{2}$

$\quad = \dfrac{k(k+1)}{2}$

$\quad = $ RHS

$P(k + 1)$ is true.

Therefore $P(n)$ is true by mathematical induction.

Question 20

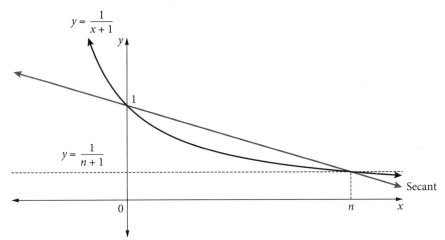

Between $x = 0$ and $x = n$ on the graph, we see that the area under the curve $y = \dfrac{1}{x+1}$ is less than the area

under the secant (a trapezium) but greater than the area under the line $y = \dfrac{1}{n+1}$ (a rectangle).

Area of rectangle $< \displaystyle\int_0^n \dfrac{1}{x+1}\,dx <$ Area of trapezium

$$\text{Area of rectangle} = n \times \dfrac{1}{n+1}$$

$$= \dfrac{n}{n+1}$$

$$\text{Area of trapezium} = \dfrac{1}{2}\left(1 + \dfrac{1}{n+1}\right)n$$

$$= \dfrac{n}{2}\left(\dfrac{n+1+1}{n+1}\right)$$

$$= \dfrac{n}{2}\left(\dfrac{n+2}{n+1}\right)$$

Area under curve:

$$\int_0^n \dfrac{1}{x+1}\,dx = \Big[\ln|x+1|\Big]_0^n$$

$$= \ln(n+1) - \ln 1 \quad n > 0$$

$$= \ln(n+1)$$

Area of rectangle $< \displaystyle\int_0^n \dfrac{1}{x+1}\,dx <$ Area of trapezium

$$\dfrac{n}{n+1} < \ln(n+1) < \dfrac{n}{2}\left(\dfrac{n+2}{n+1}\right)$$

$$n < (n+1)\ln(n+1) < \dfrac{n}{2}(n+2)$$

$$n < \ln(n+1)^{n+1} < \dfrac{n}{2}(n+2)$$

$$e^n < (n+1)^{n+1} < e^{\frac{n(n+2)}{2}} \qquad \text{QED.}$$

Question 21

a RTP: $\sqrt{ab} \le \dfrac{a+b}{2}$, where $a, b \ge 0$.

See page 9 for proof of this AM-GM inequality.

b RTP: If $1 \le x \le y$, then $x(y - x + 1) \ge y$.

Proof:

Consider the difference:

$$
\begin{aligned}
x(y - x + 1) - y &= xy - x^2 + x - y \\
&= x(y - x) - (y - x) \\
&= (y - x)(x - 1) \\
&\ge 0 \text{ because } y \ge x \text{ and } x \ge 1.
\end{aligned}
$$

So $x(y - x + 1) \ge y$.

c RTP: If $n, j \in \mathbb{N}$ and $1 \le j \le n$, then

$$\sqrt{n} \le \sqrt{j(n - j + 1)} \le \dfrac{n+1}{2}.$$

Proof: From part **a** we know that $\sqrt{ab} \le \dfrac{a+b}{2}$

So letting $a = j$ and $b = n - j + 1$, we have

$$\sqrt{j(n - j + 1)} \le \dfrac{j + (n - j + 1)}{2}$$

$$\therefore \sqrt{j(n - j + 1)} \le \dfrac{n+1}{2}, \text{ as required,}$$

for the right half of the inequality.

From part **b**, we know that $x(y - x + 1) \ge y$.

Let $x = j, y = n$:

$$j(n - j + 1) \ge n$$

Because $1 \le j \le n$, n and j are positive and $n \ge j$, so both sides of the inequality are positive.

$$\therefore \sqrt{j(n - j + 1)} \ge \sqrt{n}$$

$$\therefore \sqrt{n} \le \sqrt{j(n - j + 1)}$$

So $\sqrt{n} \le \sqrt{j(n - j + 1)} \le \dfrac{n+1}{2}.$ QED.

d RTP: for integers $n \geq 1$, $\left(\sqrt{n}\right)^n \leq n! \leq \left(\dfrac{n+1}{2}\right)^n$.

Proof:

Now $\sqrt{n} \leq \sqrt{j(n-j+1)} \leq \dfrac{n+1}{2}$ from part **c**.

So let $j = 1, 2, 3, \ldots$ and look for a pattern:

$j = 1$: $\sqrt{n} \leq \sqrt{1(n-1+1)} \leq \dfrac{n+1}{2}$

$\qquad\qquad \sqrt{n} \leq \sqrt{1(n)} \leq \dfrac{n+1}{2}$

$j = 2$: $\sqrt{n} \leq \sqrt{2(n-2+1)} \leq \dfrac{n+1}{2}$

$\qquad\qquad \sqrt{n} \leq \sqrt{2(n-1)} \leq \dfrac{n+1}{2}$

$j = 3$: $\sqrt{n} \leq \sqrt{3(n-3+1)} \leq \dfrac{n+1}{2}$

$\qquad\qquad \sqrt{n} \leq \sqrt{3(n-2)} \leq \dfrac{n+1}{2}$

$j = 4$: $\sqrt{n} \leq \sqrt{4(n-4+1)} \leq \dfrac{n+1}{2}$

$\qquad\qquad \sqrt{n} \leq \sqrt{4(n-3)} \leq \dfrac{n+1}{2}$

...

$j = n-1$: $\sqrt{n} \leq \sqrt{(n-1)(n-(n-1)+1)} \leq \dfrac{n+1}{2}$

$\qquad\qquad \sqrt{n} \leq \sqrt{(n-1)(2)} \leq \dfrac{n+1}{2}$

$j = n$: $\sqrt{n} \leq \sqrt{(n)(n-(n)+1)} \leq \dfrac{n+1}{2}$

$\qquad\qquad \sqrt{n} \leq \sqrt{(n)(1)} \leq \dfrac{n+1}{2}$

To get $\left(\sqrt{n}\right)^n$, multiply all n inequalities together:

$$\sqrt{n}\sqrt{n}\sqrt{n}\cdots\sqrt{n} \leq \sqrt{n}\sqrt{2(n-1)}\sqrt{3(n-2)}\sqrt{4(n-3)}\cdots\sqrt{(n-2)3}\sqrt{(n-1)2}\sqrt{n}$$

$$\leq \left(\dfrac{n+1}{2}\right)\left(\dfrac{n+1}{2}\right)\left(\dfrac{n+1}{2}\right)\cdots\left(\dfrac{n+1}{2}\right)$$

$$\therefore \left(\sqrt{n}\right)^n \leq \sqrt{n(n-1)(n-2)(n-3)\cdots3\cdot2\cdot1}\sqrt{2\cdot3\cdot4\cdots(n-3)(n-2)(n-1)n} \leq \left(\dfrac{n+1}{2}\right)^n$$

$$\therefore \left(\sqrt{n}\right)^n \leq \sqrt{n!}\sqrt{n!} \leq \left(\dfrac{n+1}{2}\right)^n$$

So $\left(\sqrt{n}\right)^n \leq n! \leq \left(\dfrac{n+1}{2}\right)^n$.

WORKED SOLUTIONS

HSC exam topic grid (2011–2020)

This table shows the coverage of this topic in past HSC exams by question number. The past exams can be downloaded from the NESA website (www.educationstandards.nsw.edu.au) by selecting 'Year 11 – Year 12', 'HSC exam papers'. NESA marking feedback and guidelines can also be found there.

The new Mathematics Extension 2 course was first examined in 2020. For exams before 2020, select 'Year 11 – Year 12', 'Resources archive', 'HSC exam papers archive'.

The language of proof, proof by contradiction and counterexample were introduced to the Mathematics Extension 2 course in 2020.

	The language and methods of proof	Proofs involving numbers and inequalities	Series and divisibility proofs by induction	Inequalities and other proofs by induction
2011		5(b)		3(c)
2012		**15(a)**	16(b)	
2013		14(a), 16(a)		14(b)
2014		15(a), 16(b)		
2015		15(b), 15(c)		
2016		14(c)		16(c)
2017		13(a)		16(c)
2018		15(c)	16(a)	
2019			14(c)	
2020 new course	**7**, **8**, 14(d), **15(a)**	7, 13(c), 14(d)	14(c)	

Questions in **bold** can be found in this chapter.

CHAPTER 2
3D VECTORS

3D VECTORS

Operations with vectors

- 3D vectors
- Unit vectors
- Addition and subtraction
- Multiplication by a scalar
- Magnitude of a vector
- Scalar (dot) product

Geometrical proofs in 2D and 3D

Vector equations of curves

- 3D space and coordinates
- Parametric equations of curves
- Equation of a sphere

Vector equations of lines

- Vector equation of a line $\underset{\sim}{r} = \underset{\sim}{a} + \lambda \underset{\sim}{b}$
- Testing whether a point lies on a line
- Parallel and perpendicular lines

Glossary

Cartesian equation
An equation for a line or plane in terms of x, y and z.

component form
Representation of a vector $\begin{pmatrix} a \\ b \\ c \end{pmatrix}$ in the form $a\underline{i} + b\underline{j} + c\underline{k}$,

where $\underline{i}$ is a unit vector in the x direction, $\underline{j}$ is a unit vector in the y direction and $\underline{k}$ is a unit vector in the z direction.

magnitude of a vector
The length of a vector, its numerical value. The vector $\underline{r} = x\underline{i} + y\underline{j} + z\underline{k}$ has magnitude $|\underline{r}| = \sqrt{x^2 + y^2 + z^2}$.

parallel vectors
Vectors that have the same or opposite direction.
$\underline{u} \cdot \underline{v} = \pm|\underline{u}||\underline{v}|$ if $\underline{u}$ and $\underline{v}$ are parallel.

parametric equations
A set of equations that express a set of quantities as explicit functions of a number of independent variables, known as parameters.

perpendicular vectors
Lines on a plane that have a right angle between their directions.
$\underline{u} \cdot \underline{v} = 0$ if $\underline{u}$ and $\underline{v}$ are perpendicular.

scalar
A quantity that has magnitude but no direction.

scalar (or dot) product
The product of 2 vectors as a scalar or value (not a vector).

unit vector
A vector with a magnitude of 1. The standard unit vectors are $\underline{i}$ in the x direction, $\underline{j}$ in the y direction and $\underline{k}$ in the z direction. The unit vector of $\underline{r}$ with magnitude 1 is

$$\hat{\underline{r}} = \frac{\underline{r}}{|\underline{r}|} = \frac{x\underline{i} + y\underline{j} + z\underline{k}}{\sqrt{x^2 + y^2 + z^2}}.$$

vector
A quantity with both magnitude and direction. A vector can be represented as $\underline{a}$, **a** or $\overrightarrow{AB}$.

GLOSSARY

Topic summary

Further work with vectors (MEX-V1)

V1.1 Introduction to three-dimensional vectors

3D vectors

In 3 dimensions, the unit vectors are $\underset{\sim}{i}$, $\underset{\sim}{j}$ and $\underset{\sim}{k}$ in the directions of the x-, y- and z-axes, respectively. In a 3D coordinate system, the x-y Cartesian plane 'sits on ground level' while the z-axis points upwards. Hence, the position vector of a typical point $P(x, y, z)$ from the origin is

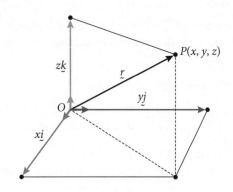

$$\underset{\sim}{r} = \overrightarrow{OP} = x\underset{\sim}{i} + y\underset{\sim}{j} + z\underset{\sim}{k}.$$

The vector $\underset{\sim}{r}$ has **magnitude** $|\underset{\sim}{r}| = \sqrt{x^2 + y^2 + z^2}$.

The **unit vector** of $\underset{\sim}{r}$, with magnitude 1, is

$$\hat{\underset{\sim}{r}} = \frac{\underset{\sim}{r}}{|\underset{\sim}{r}|} = \frac{x\underset{\sim}{i} + y\underset{\sim}{j} + z\underset{\sim}{k}}{\sqrt{x^2 + y^2 + z^2}}.$$

Example 1

Let $\underset{\sim}{u} = 2\underset{\sim}{i} - 2\underset{\sim}{j} + \underset{\sim}{k}$.

The magnitude of $\underset{\sim}{u}$:

$$\begin{aligned}
|\underset{\sim}{u}| &= \sqrt{2^2 + (-2)^2 + 1^2} \\
&= \sqrt{9} \\
&= 3
\end{aligned}$$

The unit vector of $\underset{\sim}{u}$:

$$\begin{aligned}
\hat{\underset{\sim}{u}} &= \frac{\underset{\sim}{u}}{|\underset{\sim}{u}|} \\
&= \frac{2\underset{\sim}{i} - 2\underset{\sim}{j} + \underset{\sim}{k}}{3} \quad \text{OR} \quad \frac{2}{3}\underset{\sim}{i} - \frac{2}{3}\underset{\sim}{j} + \frac{1}{3}\underset{\sim}{k}
\end{aligned}$$

Adding and subtracting vectors

Let $\underset{\sim}{u} = x_1\underset{\sim}{i} + y_1\underset{\sim}{j} + z_1\underset{\sim}{k}$ and $\underset{\sim}{v} = x_2\underset{\sim}{i} + y_2\underset{\sim}{j} + z_2\underset{\sim}{k}$.

$$\begin{aligned}
\underset{\sim}{u} + \underset{\sim}{v} &= (x_1\underset{\sim}{i} + y_1\underset{\sim}{j} + z_1\underset{\sim}{k}) + (x_2\underset{\sim}{i} + y_2\underset{\sim}{j} + z_2\underset{\sim}{k}) \\
&= (x_1 + x_2)\underset{\sim}{i} + (y_1 + y_2)\underset{\sim}{j} + (z_1 + z_2)\underset{\sim}{k}
\end{aligned}$$

$$\begin{aligned}
\underset{\sim}{u} - \underset{\sim}{v} &= (x_1\underset{\sim}{i} + y_1\underset{\sim}{j} + z_1\underset{\sim}{k}) - (x_2\underset{\sim}{i} + y_2\underset{\sim}{j} + z_2\underset{\sim}{k}) \\
&= (x_1 - x_2)\underset{\sim}{i} + (y_1 - y_2)\underset{\sim}{j} + (z_1 - z_2)\underset{\sim}{k}
\end{aligned}$$

Multiplying a vector by a scalar

$$\begin{aligned}
c\underset{\sim}{u} &= c(x_1\underset{\sim}{i} + y_1\underset{\sim}{j} + z_1\underset{\sim}{k}) \\
&= (cx_1)\underset{\sim}{i} + (cy_1)\underset{\sim}{j} + (cz_1)\underset{\sim}{k}
\end{aligned}$$

V1.2 Further operations with three-dimensional vectors

Scalar (dot) product of vectors

The **scalar** product of u and v is:

$$u \cdot v = x_1 x_2 + y_1 y_2 + z_1 z_2$$

$$u \cdot v = |u| \, |v| \cos \theta,$$

where θ is the angle between u and v.

$$u \cdot v = v \cdot u$$

$$u \cdot (v + w) = u \cdot v + u \cdot w$$

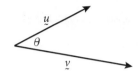

Example 2

For $u = 2i - 3j + 7k$ and $v = 4i - 5j - 3k$,

$$
\begin{aligned}
u \cdot v &= (2 \times 4) + [-3 \times (-5)] + [7 \times (-3)] \\
&= 8 + 15 - 21 \\
&= 2
\end{aligned}
$$

The angle between 2 vectors

The angle θ between the 2 vectors u and v can be found using

$$\cos \theta = \frac{u \cdot v}{|u| \, |v|}$$

Example 3

Find, correct to the nearest degree, the angle between the vectors $u = -5i + 3j - 10k$ and $v = 8i + 3k$.

Solution

$$
\begin{aligned}
u \cdot v &= (-5) \times 8 + 3 \times 0 + (-10) \times 3 \\
&= -70
\end{aligned}
$$

$$
\begin{aligned}
|u| &= \sqrt{(-5)^2 + 3^2 + (-10)^2} \\
&= \sqrt{134}
\end{aligned}
$$

$$
\begin{aligned}
|v| &= \sqrt{8^2 + 0^2 + 3^2} \\
&= \sqrt{73}
\end{aligned}
$$

$$
\begin{aligned}
\cos \theta &= \frac{u \cdot v}{|u| \, |v|} \\
&= \frac{-70}{\sqrt{134} \times \sqrt{73}} \\
&= -0.707\,757\ldots
\end{aligned}
$$

$$
\begin{aligned}
\theta &= 135.0527\ldots^\circ \\
&\approx 135^\circ
\end{aligned}
$$

Geometrical proofs using vectors

Vector properties used in geometry proofs

Scalar (dot) product: $\underset{\sim}{u} \cdot \underset{\sim}{v} = |\underset{\sim}{u}||\underset{\sim}{v}|\cos\theta$, when the angle between 2 vectors is required.

Parallel vectors: If $\underset{\sim}{u}$ and $\underset{\sim}{v}$ are parallel, $\theta = 0°$ or $180°$, so $\underset{\sim}{u} \cdot \underset{\sim}{v} = \pm|\underset{\sim}{u}||\underset{\sim}{v}|$.

$\underset{\sim}{u} \cdot \underset{\sim}{v} = |\underset{\sim}{u}||\underset{\sim}{v}|$ if the vectors are in the same direction and $\underset{\sim}{u} \cdot \underset{\sim}{v} = -|\underset{\sim}{u}||\underset{\sim}{v}|$ if they are in opposite directions.

A vector parallel to $x\underset{\sim}{i} + y\underset{\sim}{j} + z\underset{\sim}{k}$ is $cx\underset{\sim}{i} + cy\underset{\sim}{j} + cz\underset{\sim}{k}$, where c is a scalar.

Write and solve $\underset{\sim}{u} \cdot \underset{\sim}{v} = |\underset{\sim}{u}||\underset{\sim}{v}|$.

Perpendicular vectors: If $\underset{\sim}{u}$ and $\underset{\sim}{v}$ are perpendicular, $\theta = 90°$, so $\underset{\sim}{u} \cdot \underset{\sim}{v} = 0$.

Write and solve $\underset{\sim}{u} \cdot \underset{\sim}{v} = 0$.

Midpoint of vectors:

$$\frac{\underset{\sim}{u}}{2} \qquad \text{for a single vector}$$

$$\frac{\underset{\sim}{u} + \underset{\sim}{v}}{2} \qquad \text{for the sum of 2 vectors}$$

$$\frac{\underset{\sim}{u} - \underset{\sim}{v}}{2} \qquad \text{for the vector from } \underset{\sim}{u} \text{ to the midpoint between } \underset{\sim}{u} \text{ and } \underset{\sim}{v}.$$

Example 4

One pair of opposite sides of a quadrilateral are parallel and equal in length.

Show that the other 2 sides are parallel and equal.

Solution

$\overrightarrow{AD}$ is the vector from A to D and is expressed as $\underset{\sim}{d} - \underset{\sim}{a}$.

Let the sides AD and BC be parallel and equal, so vectors $\underset{\sim}{d} - \underset{\sim}{a}$ and $\underset{\sim}{c} - \underset{\sim}{b}$ are in the same direction and equal in length.

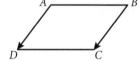

We need to show that the other 2 sides, $\underset{\sim}{b} - \underset{\sim}{a}$ and $\underset{\sim}{c} - \underset{\sim}{d}$, are also parallel and equal in length.

Since $\underset{\sim}{d} - \underset{\sim}{a} = \underset{\sim}{c} - \underset{\sim}{b}$:

$$\underset{\sim}{d} - \underset{\sim}{a} + \underset{\sim}{b} = \underset{\sim}{c}$$
$$-\underset{\sim}{a} + \underset{\sim}{b} = \underset{\sim}{c} - \underset{\sim}{d}$$
$$\therefore \underset{\sim}{b} - \underset{\sim}{a} = \underset{\sim}{c} - \underset{\sim}{d}, \text{ as required.}$$

So vectors $\overrightarrow{AB}$ and $\overrightarrow{DC}$ are equal in size and direction.

Therefore, they are parallel and equal in length.

Example 5   ©NESA   2020 HSC EXAM, QUESTION 15(b)

The point C divides the interval AB so that $\dfrac{CB}{AC} = \dfrac{m}{n}$. The position vectors of A and B are $\underset{\sim}{a}$ and $\underset{\sim}{b}$ respectively, as shown in the diagram.

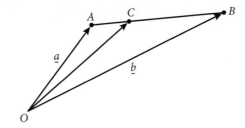

a   Show that $\overrightarrow{AC} = \dfrac{n}{m+n}(\underset{\sim}{b} - \underset{\sim}{a})$.

b   Prove that $\overrightarrow{OC} = \dfrac{m}{m+n}\,\underset{\sim}{a} + \dfrac{n}{m+n}\,\underset{\sim}{b}$

Solution

a $\dfrac{CB}{AC} = \dfrac{m}{n}$

So AB is divided into $m + n$ parts.

Then

$\overrightarrow{AB} = \overrightarrow{AC} + \overrightarrow{CB}$

$\qquad = \left(\dfrac{n}{m+n}\right)\overrightarrow{AB} + \left(\dfrac{m}{m+n}\right)\overrightarrow{AB}.$

Now,

$\underset{\sim}{a} + \overrightarrow{AB} = \underset{\sim}{b}$

$\qquad \overrightarrow{AB} = \underset{\sim}{b} - \underset{\sim}{a}$

so

$\overrightarrow{AC} = \overrightarrow{AB}\left(\dfrac{n}{m+n}\right)$

$\qquad = (\underset{\sim}{b} - \underset{\sim}{a})\left(\dfrac{n}{m+n}\right).$

b $\overrightarrow{OC} = \underset{\sim}{a} + \overrightarrow{AC}$

$\qquad = \dfrac{\underset{\sim}{a}(m+n)}{m+n} + \left(\dfrac{n}{m+n}\right)(\underset{\sim}{b} - \underset{\sim}{a})$ (from part **a**)

$\qquad = \dfrac{\underset{\sim}{a}m + \underset{\sim}{a}n + \underset{\sim}{b}n - \underset{\sim}{a}n}{m+n}$

$\qquad = \dfrac{\underset{\sim}{a}m + \underset{\sim}{b}n}{m+n}$

$\qquad = \left(\dfrac{m}{m+n}\right)\underset{\sim}{a} + \left(\dfrac{n}{m+n}\right)\underset{\sim}{b}$

V1.3 Vectors and vector equations of lines

3D space

In three-dimensional space, every point $P(x, y, z)$ can be determined from the origin by the position vector $\overrightarrow{OP}$, which can be represented by the vector $\underset{\sim}{r} = x\underset{\sim}{i} + y\underset{\sim}{j} + z\underset{\sim}{k}$.

Example 6

Find the distance between the points $A(2, 3, -1)$ and $B(3, -2, 1)$.

Solution

$$\text{Distance} = \sqrt{(3-2)^2 + (-2-3)^2 + (1-[-1])^2}$$
$$= \sqrt{30}$$

Example 7

Find the angle between the vectors $\underset{\sim}{u}$ and $\underset{\sim}{v}$ if $\underset{\sim}{u} = \begin{pmatrix} 2 \\ 3 \\ -1 \end{pmatrix}$ and $\underset{\sim}{v} = \begin{pmatrix} 3 \\ -2 \\ 1 \end{pmatrix}$, correct to one decimal place.

Solution

$$\cos\theta = \frac{\underset{\sim}{u} \cdot \underset{\sim}{v}}{|\underset{\sim}{u}||\underset{\sim}{v}|}$$

$$\underset{\sim}{u} \cdot \underset{\sim}{v} = 2(3) + 3(-2) + (-1)1 = -1$$

$$|\underset{\sim}{u}| = \sqrt{2^2 + 3^2 + (-1)^2} = \sqrt{14}$$

$$|\underset{\sim}{v}| = \sqrt{3^2 + (-2)^2 + 1^2} = \sqrt{14}$$

$$|\underset{\sim}{u}||\underset{\sim}{v}| = \sqrt{14} \times \sqrt{14}$$
$$= 14$$

$$\therefore \cos\theta = -\frac{1}{14}$$
$$\theta = 94.0960\ldots$$
$$\approx 94.1°$$

Hence, the angle between the vectors is $94.1°$.

Parametric equations of a curve

A convenient way to describe a curve in 3D space is to provide a vector that 'points to' every point on the curve as the parameter (t) varies, for example, $\begin{pmatrix} 1 + 3t \\ 2 - 2t \\ 3 + t \end{pmatrix}$.

We call $\begin{pmatrix} f(t) \\ g(t) \\ h(t) \end{pmatrix}$ a vector function and $x = f(t)$, $y = g(t)$ and $z = h(t)$ are called the **parametric equations** of the curve. We often think of the parameter t as time and thus the parametric equations indicate the position of an object in 3D space at any time.

When graphing lines and curves in 3D space, draw the axes as shown.

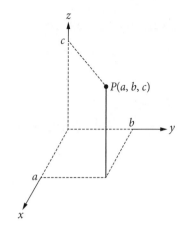

This is the graph of $\begin{pmatrix} t \\ t \\ t \end{pmatrix}$ for $0 \le t \le 10$, a straight line that goes

through $(0,0,0)$ and $(10,10,10)$ with parametric equations

$$x = t, y = t, z = t.$$

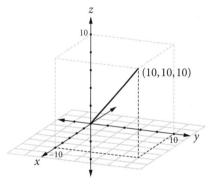

Example 8

Draw the curve produced by $x = \cos t$, $y = \sin t$ and $z = t$, $\begin{pmatrix} \cos t \\ \sin t \\ t \end{pmatrix}$ for $0 \le t \le 2\pi$.

Solution

The parametric equations are:

$$x = \cos t, y = \sin t \text{ and } z = t.$$

As t increases from 0, $x = \cos t$, $y = \sin t$ moves along the unit circle with centre $(0,0)$ on the x-y plane, starting at $(1,0)$ and then moving in an anticlockwise direction. For the z-coordinate, $z = t$ so the unit circle is moving steadily away from the x-y plane, creating a rising curve.

See the table of values for the (x, y, z) coordinates.

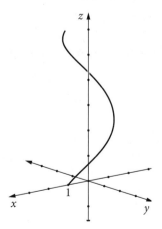

t	$x = \cos t$	$y = \sin t$	$z = t$
0	1	0	0
$\dfrac{\pi}{4}$	$\dfrac{1}{\sqrt{2}}$	$\dfrac{1}{\sqrt{2}}$	$\dfrac{\pi}{4}$
$\dfrac{\pi}{2}$	0	1	$\dfrac{\pi}{2}$
$\dfrac{3\pi}{4}$	$-\dfrac{1}{\sqrt{2}}$	$\dfrac{1}{\sqrt{2}}$	$\dfrac{3\pi}{4}$
π	-1	0	π
$\dfrac{5\pi}{4}$	$-\dfrac{1}{\sqrt{2}}$	$-\dfrac{1}{\sqrt{2}}$	$\dfrac{5\pi}{4}$
$\dfrac{3\pi}{2}$	0	-1	$\dfrac{3\pi}{2}$
$\dfrac{7\pi}{4}$	$\dfrac{1}{\sqrt{2}}$	$-\dfrac{1}{\sqrt{2}}$	$\dfrac{7\pi}{4}$
2π	1	0	2π

The curve is in the shape of a helix beginning at $(1,0,0)$ and ending at $(1,0,2\pi)$, directly above its starting point, after one revolution of the circle.

Equation of a sphere

In 3 dimensions, the Cartesian equation

$$x^2 + y^2 + z^2 = r^2$$

represents a sphere with centre $(0,0,0)$ and radius r.

More generally, the equation

$$(x - a)^2 + (y - b)^2 + (z - c)^2 = r^2$$

represents a sphere with centre (a, b, c) and radius r.

Example 9

Describe the sphere represented by each equation.

a $(x - 1)^2 + (y + 2)^2 + (z - 3)^2 = 25$

b $x^2 + 4x + y^2 + 6y + z^2 - 2z = 2$

Solution

a This is a sphere with centre $(1, -2, 3)$ and radius $\sqrt{25} = 5$.

b Completing the square for x, y and z:

$$x^2 + 4x + 4 + y^2 + 6y + 9 + z^2 - 2z + 1 = 2 + (4 + 9 + 1)$$
$$(x + 2)^2 + (y + 3)^2 + (z - 1)^2 = 16$$

This is a sphere with centre $(-2, -3, 1)$ and radius $\sqrt{16} = 4$.

Vector equation of a straight line

In order to specify a straight line, we need to know 2 things: a point through which the line passes, and the line's direction.

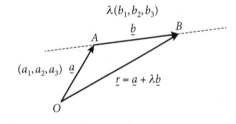

The vector equation of a straight line through points A and B is $\underset{\sim}{r} = \underset{\sim}{a} + \lambda \underset{\sim}{b}$, where $\underset{\sim}{a} = \overrightarrow{OA}$ is the vector from the origin to A, $\underset{\sim}{b} = \overrightarrow{AB}$ is the direction of the required line, λ is a parameter and $\underset{\sim}{r} = \overrightarrow{OR}$, where R is a point on AB.

For example, a line through the point $(2, -1, 0)$, parallel to the vector $-2\underset{\sim}{i} + \underset{\sim}{j} + \underset{\sim}{k}$, has equation:

$$\underset{\sim}{r} = \underset{\sim}{a} + \lambda \underset{\sim}{b}$$
$$= (2\underset{\sim}{i} - \underset{\sim}{j} + 0\underset{\sim}{k}) + \lambda(-2\underset{\sim}{i} + \underset{\sim}{j} + \underset{\sim}{k})$$
$$= (2\underset{\sim}{i} - \underset{\sim}{j}) + \lambda(-2\underset{\sim}{i} + \underset{\sim}{j} + \underset{\sim}{k})$$

$\underset{\sim}{b}$ is the direction vector in the 3D line $\underset{\sim}{r} = \underset{\sim}{a} + \lambda \underset{\sim}{b}$, similar to the gradient m in the equation of the 2D line $y = mx + c$.

The **Cartesian form of the straight line** is found by equating the different expressions for λ:

$$\lambda = \frac{x - x_1}{x_2 - x_1} = \frac{y - y_1}{y_2 - y_1} = \frac{z - z_1}{z_2 - z_1}$$

Example 10

A line passes through the points $A(-2, 1, 5)$ and $B(6, 3, -4)$.

a Write a vector equation of the line.

b Write parametric equations for the line.

c Determine if the point $C(-10, -1, 14)$ lies on the line.

Solution

a First find the direction vector, $\underset{\sim}{b}$:

$$\underset{\sim}{b} = \overrightarrow{OB} - \overrightarrow{OA}$$

$$= \begin{pmatrix} 6 \\ 3 \\ -4 \end{pmatrix} - \begin{pmatrix} -2 \\ 1 \\ 5 \end{pmatrix}$$

$$= \begin{pmatrix} 8 \\ 2 \\ -9 \end{pmatrix}$$

A vector equation for the line is:

$$\underset{\sim}{r} = \underset{\sim}{a} + \lambda\underset{\sim}{b}$$

$$\underset{\sim}{r} = \begin{pmatrix} x \\ y \\ z \end{pmatrix} = \begin{pmatrix} -2 \\ 1 \\ 5 \end{pmatrix} + \lambda\begin{pmatrix} 8 \\ 2 \\ -9 \end{pmatrix} \quad \text{OR} \quad \begin{pmatrix} 6 \\ 3 \\ -4 \end{pmatrix} + \lambda\begin{pmatrix} 8 \\ 2 \\ -9 \end{pmatrix}$$

b The corresponding parametric equations are:

$$x = -2 + 8\lambda$$
$$y = 1 + 2\lambda$$
$$z = 5 - 9\lambda$$

c If the point lies on the line, then substitute the coordinates $(-10, -1, 14)$ for x, y, z and check λ:

$$-10 = -2 + 8\lambda, \quad \text{so } \lambda = -1$$
$$-1 = 1 + 2\lambda, \quad \text{so } \lambda = -1$$
$$14 = 5 - 9\lambda, \quad \text{so } \lambda = -1$$

$\lambda = -1$ for all 3 coordinates.

So C does lie on the line joining A to B.

Example 11

Determine whether the lines $\begin{pmatrix} 1 \\ 1 \\ 1 \end{pmatrix} + \lambda_1 \begin{pmatrix} 1 \\ 2 \\ -1 \end{pmatrix}$ and $\begin{pmatrix} 3 \\ 2 \\ 1 \end{pmatrix} + \lambda_1 \begin{pmatrix} -1 \\ -5 \\ 3 \end{pmatrix}$ are parallel, intersect or neither.

Solution

In the 2 equations, we can see the direction vectors $\begin{pmatrix} 1 \\ 2 \\ -1 \end{pmatrix}$ and $\begin{pmatrix} -1 \\ -5 \\ 3 \end{pmatrix}$ are not scalar multiples of each other, and so they are not parallel.

If they intersect there must be 2 values a and b such that:

$$\begin{pmatrix} 1 \\ 1 \\ 1 \end{pmatrix} + a \begin{pmatrix} 1 \\ 2 \\ -1 \end{pmatrix} = \begin{pmatrix} 3 \\ 2 \\ 1 \end{pmatrix} + b \begin{pmatrix} -1 \\ -5 \\ 3 \end{pmatrix}.$$

Therefore,

$$x = 1 + a = 3 - b$$

$$y = 1 + 2a = 2 - 5b$$

$$z = 1 - a = 1 + 3b.$$

Solving any 2 equations above simultaneously, we get $a = 3$ and $b = -1$.

These values satisfy all 3 equations and so tell us that the 2 lines intersect at $(4, 7, -2)$.

Example 12 ©NESA 2020 HSC EXAM, QUESTION 3

What is the Cartesian equation of the line $\underset{\sim}{r} = \begin{pmatrix} 1 \\ 3 \end{pmatrix} + \lambda \begin{pmatrix} -2 \\ 4 \end{pmatrix}$?

A $2y + x = 7$

B $y - 2x = -5$

C $y + 2x = 5$

D $2y - x = -1$

Solution

$x = 1 - 2\lambda$ [1]

$y = 3 + 4\lambda$ [2]

Eliminating the parameter λ to find y in terms of x:

$2 \times [1]$:
$2x = 2 - 4\lambda$ [3]

$[2] + [3]$:
$y + 2x = 5$ (option C)

Parallel and perpendicular lines

The scalar product can be used to determine the angle between 2 vectors. In particular, we are interested in whether the angle between the vectors is $0°$ or $180°$ (parallel), or $90°$ or $270°$ (perpendicular).

$$\underset{\sim}{u} \cdot \underset{\sim}{v} = |\underset{\sim}{u}||\underset{\sim}{v}|\cos\theta$$

$$\text{OR } \cos\theta = \frac{\underset{\sim}{u} \cdot \underset{\sim}{v}}{|\underset{\sim}{u}||\underset{\sim}{v}|}$$

> **Hint**
> If $\underset{\sim}{u} \cdot \underset{\sim}{v} = 0$, then $\cos\theta = 0$ and the vectors are perpendicular.
> If $\underset{\sim}{u} \cdot \underset{\sim}{v} = \pm|\underset{\sim}{u}||\underset{\sim}{v}|$, then $\cos\theta = \pm1$ and the vectors are parallel.

TOPIC SUMMARY

Example 13

Show that $2\underset{\sim}{i} + 3\underset{\sim}{j} - 2\underset{\sim}{k}$ and $\underset{\sim}{i} - 2\underset{\sim}{j} - 2\underset{\sim}{k}$ are perpendicular vectors.

Solution

$$\underset{\sim}{u} \cdot \underset{\sim}{v} = x_1 x_2 + y_1 y_2 + z_1 z_2$$
$$= 2 \times 1 + 3(-2) + (-2)(-2)$$
$$= 0$$

$$\cos\theta = \frac{\underset{\sim}{u} \cdot \underset{\sim}{v}}{|\underset{\sim}{u}||\underset{\sim}{v}|} = \frac{0}{|\underset{\sim}{u}||\underset{\sim}{v}|}$$

$$\therefore \cos\theta = 0$$
$$\text{So } \theta = 90°$$

Hence, the vectors are perpendicular.

Example 14

Show that $2\underset{\sim}{i} + 3\underset{\sim}{j} - 2\underset{\sim}{k}$ and $-4\underset{\sim}{i} - 6\underset{\sim}{j} + 4\underset{\sim}{k}$ are parallel vectors.

Solution

$$\underset{\sim}{u} \cdot \underset{\sim}{v} = 2(-4) + 3(-6) + 4(-2)$$
$$= -34$$

$$|\underset{\sim}{u}| = \sqrt{2^2 + 3^2 + (-2)^2}$$
$$= \sqrt{17}$$

$$|\underset{\sim}{v}| = \sqrt{(-4)^2 + (-6)^2 + 4^2}$$
$$= \sqrt{68}$$

$$\cos\theta = \frac{\underset{\sim}{u} \cdot \underset{\sim}{v}}{|\underset{\sim}{u}||\underset{\sim}{v}|}$$
$$= \frac{-34}{\sqrt{17}\sqrt{68}}$$
$$= \frac{-34}{\sqrt{1156}}$$
$$= \frac{-34}{34}$$
$$= -1$$
$$\theta = 180°$$

Hence, the vectors are parallel, but opposite in direction.

OR

Notice that $\underset{\sim}{v}$ is a scalar multiple of $\underset{\sim}{u}$:

$$\underset{\sim}{v} = -4\underset{\sim}{i} - 6\underset{\sim}{j} + 4\underset{\sim}{k}$$
$$= -2(2\underset{\sim}{i} + 3\underset{\sim}{j} - 2\underset{\sim}{k})$$
$$= -2\underset{\sim}{u}$$

One vector is a scalar multiple of the other, so they are parallel.

Practice set 1

Multiple-choice questions

Solutions start on page 56.

Question 1 ☉●●
Find the length of the vector $2\underset{\sim}{i} - 3\underset{\sim}{j} + 6\underset{\sim}{k}$.

A 5   **B** 6   **C** 7   **D** 49

Question 2 ☉●●
What is the distance of $P(4, -2, 4)$ from the origin?

A $\sqrt{20}$ **B** $\sqrt{32}$ **C** 6   **D** 36

Question 3 ☉●●
What is the Cartesian equation of the line $\underset{\sim}{r} = \begin{pmatrix} 2 \\ 5 \end{pmatrix} + \lambda \begin{pmatrix} -1 \\ 3 \end{pmatrix}$?

A $y + 2x = 8$ **B** $y = 11 - 3x$ **C** $x + y = 7$ **D** $y = -x + 10$

Question 4 ☉●●
Which vector below is NOT parallel to $-2\underset{\sim}{i} + 4\underset{\sim}{j} - 3\underset{\sim}{k}$?

A $2\underset{\sim}{i} - 4\underset{\sim}{j} + 3\underset{\sim}{k}$ **B** $-2\underset{\sim}{i} + 4\underset{\sim}{j} + 3\underset{\sim}{k}$ **C** $8\underset{\sim}{i} - 16\underset{\sim}{j} + 12\underset{\sim}{k}$ **D** $-4\underset{\sim}{i} + 8\underset{\sim}{j} - 6\underset{\sim}{k}$

Question 5 ☉●●
Find the direction vector for the line $\dfrac{x - 1}{2} = \dfrac{y + 3}{-1} = \dfrac{z - 2}{5}$.

A $\underset{\sim}{i} - 3\underset{\sim}{j} + 2\underset{\sim}{k}$ **B** $-\underset{\sim}{i} + 3\underset{\sim}{j} - 2\underset{\sim}{k}$ **C** $2\underset{\sim}{i} - \underset{\sim}{j} + 5\underset{\sim}{k}$ **D** $-5\underset{\sim}{i} + \underset{\sim}{j} + 2\underset{\sim}{k}$

Question 6 ☉●●
Find the vector equation of the line passing through the points $(1, -3, 2)$ and $(-1, 2, -2)$.

A $\underset{\sim}{i} - 3\underset{\sim}{j} + 2\underset{\sim}{k}$

B $(-\underset{\sim}{i} + 2\underset{\sim}{j} - 2\underset{\sim}{k}) + \lambda(2\underset{\sim}{i} - 5\underset{\sim}{j} + 4\underset{\sim}{k})$

C $-\underset{\sim}{i} + 2\underset{\sim}{j} - 2\underset{\sim}{k}$

D $(-2\underset{\sim}{i} + 5\underset{\sim}{j} - 4\underset{\sim}{k}) + \lambda(-\underset{\sim}{i} + 2\underset{\sim}{j} - 2\underset{\sim}{k})$

Question 7 ●●☉
Which of the following points is NOT on the line passing through $(2, -1, 3)$ and $(4, 2, -1)$?

A $(4, 2, -1)$ **B** $(0, -4, 7)$ **C** $(6, 5, -5)$ **D** $(4, -4, 1)$

Question 8 ☉●●
Which of the following defines a line perpendicular to $\begin{pmatrix} 2 \\ -1 \\ 3 \end{pmatrix} + \lambda \begin{pmatrix} -1 \\ 3 \\ 2 \end{pmatrix}$?

A $\dfrac{x - 2}{2} = \dfrac{y + 1}{4} = \dfrac{z - 3}{-5}$

B $\begin{pmatrix} 2 \\ -1 \\ 3 \end{pmatrix} + \lambda \begin{pmatrix} 1 \\ 3 \\ 2 \end{pmatrix}$

C $(-\underset{\sim}{i} + \underset{\sim}{j} - \underset{\sim}{k}) + \lambda(-\underset{\sim}{i} + 3\underset{\sim}{j} + 2\underset{\sim}{k})$

D $\begin{pmatrix} -1 \\ 3 \\ 2 \end{pmatrix}$

Question 9 ●●

Consider the vectors $a = mi + j$ and $b = i + mj$, where $m \in \mathbb{R}$.

If the acute angle between a and b is 60°, then find the value of m.

A 1   **B** $2 \pm \sqrt{3}$ **C** $2, \dfrac{1}{2}$ **D** $\dfrac{2\sqrt{3} \pm 3}{3}$

Question 10 ●●

In the parallelogram on the right, $|a| = 2|b|$.

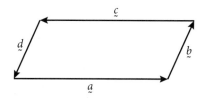

Which of the following statements is true?

A $a = 2b$ **B** $a + b = c + d$

C $a + c = 0$ **D** $a - b = c - d$

Question 11 ●●

$p = i + 2j - k$ and $q = mi - 2j$, where m is a real constant.

If vector $p - q$ is perpendicular to vector p, then find the value(s) of m.

A 0   **B** 2   **C** 0 or 2   **D** 10

Question 12 ●●

Find the values of r if the vectors $2i - rj + 5k$ and $3i - rj + rk$ are perpendicular.

A −3 or −2   **B** 3 or 2   **C** −6 or 1   **D** 6 or −1

Question 13 ●●

The vectors $a = 3i + nj + 4k$ and $b = n^2i + j - k$ are perpendicular. Find the values of n.

A $\dfrac{4}{3}$ or 1   **B** $\dfrac{4}{3}$ or −1   **C** $-\dfrac{4}{3}$ or −1   **D** $-\dfrac{4}{3}$ or 1

Question 14 ●●

A rhombus has vectors a and b, as shown.

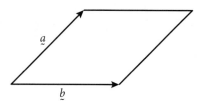

Which of the following equations must be true?

A $a \cdot b = 0$ **B** $a = b$

C $(a + b) \cdot (a - b) = 0$ **D** $|a + b| = |a - b|$

Question 15 ●●○

The position vector of a particle at time t seconds, $t \geq 0$, is given by $\underset{\sim}{r}(t) = t^2\underset{\sim}{i} - 8\sqrt{t}\underset{\sim}{j} + \dfrac{16}{t}\underset{\sim}{k}$.

Find the velocity vector when $t = 4$.

A $16\underset{\sim}{i} - 16\underset{\sim}{j} + 4\underset{\sim}{k}$ **B** $8\underset{\sim}{i} - 2\underset{\sim}{j} - \underset{\sim}{k}$ **C** $2\underset{\sim}{i} + \dfrac{1}{4}\underset{\sim}{j} - \dfrac{1}{2}\underset{\sim}{k}$ **D** $8\underset{\sim}{i} - 2\underset{\sim}{j} + \underset{\sim}{k}$

Question 16 ●●●

Which of the following lines is a tangent to the sphere described by $(x - 2)^2 + y^2 + (z + 1)^2 = 25$?

A $\underset{\sim}{r} = (2\underset{\sim}{i} + 4\underset{\sim}{j} + 2\underset{\sim}{k}) + \lambda(4\underset{\sim}{j} + 3\underset{\sim}{k})$ **B** $\underset{\sim}{r} = (4\underset{\sim}{j} + 3\underset{\sim}{k}) + \lambda(2\underset{\sim}{i} + 4\underset{\sim}{j} + 2\underset{\sim}{k})$

C $\underset{\sim}{r} = (5\underset{\sim}{i} + 3\underset{\sim}{k}) + \lambda(-4\underset{\sim}{i} + 3\underset{\sim}{k})$ **D** $\underset{\sim}{r} = (-4\underset{\sim}{i} + 3\underset{\sim}{k}) + \lambda(5\underset{\sim}{i} + 3\underset{\sim}{k})$

Question 17 ●●●

A body has displacement of $-2\underset{\sim}{i} + 3\underset{\sim}{j}$ metres at a particular time. The body moves with constant velocity and 2 seconds later has displacement of $4\underset{\sim}{i} - 5\underset{\sim}{j}$ metres.

What is the velocity of the body?

A $4\underset{\sim}{i} - 3\underset{\sim}{j}$ **B** $3\underset{\sim}{i} - 4\underset{\sim}{j}$ **C** $4\underset{\sim}{i} + 3\underset{\sim}{j}$ **D** $3\underset{\sim}{i} + 4\underset{\sim}{j}$

Question 18 ●●●

Given that $\underset{\sim}{A} + \underset{\sim}{B} = \underset{\sim}{C}$ and $\left|\underset{\sim}{A}\right|^2 + \left|\underset{\sim}{B}\right|^2 = \left|\underset{\sim}{C}\right|^2$, find the angle between $\underset{\sim}{A}$ and $\underset{\sim}{B}$.

A 0 **B** $\dfrac{\pi}{4}$ **C** $\dfrac{\pi}{2}$ **D** π

Question 19 ●●●

Find the shortest distance between the vector lines $\underset{\sim}{r_1} = (\underset{\sim}{i} + 2\underset{\sim}{j} - \underset{\sim}{k}) + \lambda(2\underset{\sim}{i} + \underset{\sim}{j} + \underset{\sim}{k})$ and $\underset{\sim}{r_2} = (3\underset{\sim}{i} - \underset{\sim}{j} + 2\underset{\sim}{k}) + \mu(\underset{\sim}{i} + \underset{\sim}{j} + \underset{\sim}{k})$, where $\mu > 0$ and $\lambda > 0$.

A $\dfrac{\sqrt{3}}{2}$ **B** $\dfrac{\sqrt{2}}{3}$ **C** $2\sqrt{3}$ **D** $3\sqrt{2}$

Question 20 ●●●

A, B and C are 3 collinear points with position vectors $\underset{\sim}{a}$, $\underset{\sim}{b}$ and $\underset{\sim}{c}$, respectively.

B lies between A and C, and $3\left|\overrightarrow{BC}\right| = \left|\overrightarrow{AB}\right|$.

Find vector $\underset{\sim}{c}$ in terms of $\underset{\sim}{a}$ and $\underset{\sim}{b}$.

A $\dfrac{4}{3}\underset{\sim}{b} - \dfrac{1}{3}\underset{\sim}{a}$ **B** $\dfrac{4}{3}\underset{\sim}{a} - \dfrac{1}{3}\underset{\sim}{b}$ **C** $\dfrac{1}{3}\underset{\sim}{a} - \dfrac{4}{3}\underset{\sim}{b}$ **D** $\dfrac{1}{3}\underset{\sim}{b} - \dfrac{4}{3}\underset{\sim}{a}$

Practice set 2

Short-answer questions

Solutions start on page 58.

Question 1 (2 marks) ●●●

a Find the unit vector for $2\underset{\sim}{i} - 3\underset{\sim}{j} + 6\underset{\sim}{k}$.   1 mark

b Find the length of the vector $4\underset{\sim}{i} - 2\underset{\sim}{j} + 4\underset{\sim}{k}$.   1 mark

Question 2 (2 marks) ●●●

Find the unit vector that has the same direction as $\underset{\sim}{v} = \begin{pmatrix} 4 \\ -2 \\ -4 \end{pmatrix}$.   2 marks

Question 3 (6 marks) ●●●

a Determine the vector $\overrightarrow{AB}$ given that $\overrightarrow{OA} = 2\underset{\sim}{i} - 4\underset{\sim}{j} + 3\underset{\sim}{k}$ and $\overrightarrow{OB} = -\underset{\sim}{i} - 4\underset{\sim}{j} + 5\underset{\sim}{k}$.   2 marks

b Determine the vector $\overrightarrow{OB}$ given that $\overrightarrow{OA} = \underset{\sim}{i} - 4\underset{\sim}{j} - 3\underset{\sim}{k}$ and $\overrightarrow{AB} = -2\underset{\sim}{i} + 4\underset{\sim}{j} - \underset{\sim}{k}$.   2 marks

c Determine the vector $\overrightarrow{AO}$ given that $\overrightarrow{OB} = 3\underset{\sim}{i} + 4\underset{\sim}{j} + \underset{\sim}{k}$ and $\overrightarrow{AB} = \underset{\sim}{i} - 3\underset{\sim}{j} + \underset{\sim}{k}$.   2 marks

Question 4 (2 marks) ●●●

Consider the 2 vectors $\underset{\sim}{u} = m\underset{\sim}{i} - 4\underset{\sim}{j} + \underset{\sim}{k}$ and $\underset{\sim}{v} = -\underset{\sim}{i} + 2\underset{\sim}{j} + 4\underset{\sim}{k}$.

For what values of m are $\underset{\sim}{u} - \underset{\sim}{v}$ and $\underset{\sim}{u} + \underset{\sim}{v}$ perpendicular?   2 marks

Question 5 (2 marks) ●●●

The Cartesian equation of a line is $\dfrac{x - 2}{-3} = \dfrac{y + 5}{4} = \dfrac{z - 1}{1}$.

Find the vector equation of this line.   2 marks

Question 6 (4 marks) ●●●

Show that the points described by the vectors $3\underset{\sim}{i} + 4\underset{\sim}{j} - 2\underset{\sim}{k}$, $7\underset{\sim}{i} + 8\underset{\sim}{j} - 8\underset{\sim}{k}$ and $13\underset{\sim}{i} + 14\underset{\sim}{j} - 17\underset{\sim}{k}$ are collinear.   4 marks

Question 7 (2 marks) ●●●

The position vectors of the points P, Q and R are $-2\underset{\sim}{i} + \underset{\sim}{j} - \underset{\sim}{k}$, $-4\underset{\sim}{i} + 2\underset{\sim}{j} + 2\underset{\sim}{k}$ and $6\underset{\sim}{i} - 3\underset{\sim}{j} - 13\underset{\sim}{k}$, respectively.

If $\overrightarrow{PQ} = \lambda \overrightarrow{PR}$, find the value of λ.   2 marks

Question 8 (2 marks) ●●●

Find, correct to the nearest degree, the angle between the line vectors $\underset{\sim}{a} = 2\underset{\sim}{i} + \underset{\sim}{j} - 2\underset{\sim}{k}$ and $\underset{\sim}{b} = 3\underset{\sim}{i} - 2\underset{\sim}{j} - \underset{\sim}{k}$.   2 marks

Question 9 (3 marks) ●●●

Find a general vector that is perpendicular to $\underset{\sim}{u} = 3\underset{\sim}{i} - 4\underset{\sim}{j} + 5\underset{\sim}{k}$.   3 marks

Question 10 (3 marks) ●●●

The equations of 2 intersecting lines are $\underset{\sim}{r} = (-2\underset{\sim}{i} + \underset{\sim}{j} - 4\underset{\sim}{k}) + \gamma(-\underset{\sim}{i} + 2\underset{\sim}{j} + 4\underset{\sim}{k})$ and $\underset{\sim}{r} = (\underset{\sim}{i} + 2\underset{\sim}{j} + 3\underset{\sim}{k}) + \mu(-m\underset{\sim}{i} + 2\underset{\sim}{j} - 4\underset{\sim}{k})$, where γ and μ are parameters and m is a constant.

If these 2 lines are perpendicular, what is the value of m?   3 marks

Question 11 (2 marks) ⬤⬤⬜

Find the point of intersection of the 2 lines

2 marks

$$\frac{x+5}{3} = \frac{y-2}{2} = \frac{z+7}{6} \quad \text{and} \quad \frac{x}{1} = \frac{y+6}{-5} = \frac{z+3}{-1}.$$

Question 12 (2 marks) ⬤⬤⬜

Find the point of intersection of the 2 lines

2 marks

$$\underset{\sim}{r} = \begin{pmatrix} 2 \\ 1 \\ -2 \end{pmatrix} + \gamma_1 \begin{pmatrix} 6 \\ -8 \\ 10 \end{pmatrix} \text{ and } \underset{\sim}{r} = \begin{pmatrix} -1 \\ -10 \\ 5 \end{pmatrix} + \gamma_2 \begin{pmatrix} 6 \\ 7 \\ -2 \end{pmatrix}.$$

Question 13 (3 marks) ⬤⬤⬜

A sphere has equation $x^2 - 2x + y^2 + 4y + z^2 - 11 = 0$.

What is the centre and radius of this sphere?

3 marks

Question 14 (3 marks) ⬤⬤⬤

Determine the Cartesian equation that would describe the path of the object that has position vector $\underset{\sim}{r} = t\underset{\sim}{i} - \frac{1}{2}t^2\underset{\sim}{j}$, $t \geq 0$ and sketch the path.

3 marks

Question 15 (5 marks) ⬤⬤⬤

The points P and Q have position vectors given by

$$\overrightarrow{OP} = \begin{pmatrix} -2 \\ 2 \\ 1 \end{pmatrix} \text{ and } \overrightarrow{OQ} = \begin{pmatrix} 0 \\ 3 \\ 4 \end{pmatrix}.$$

a   Find an expression for the vector $\overrightarrow{PQ}$ in the form $a\underset{\sim}{i} + b\underset{\sim}{j} + c\underset{\sim}{k}$.

1 mark

b   Show that the cosine of the angle between the vectors $\overrightarrow{OP}$ and $\overrightarrow{OQ}$ is $\frac{2}{3}$.

2 marks

c   Hence, find the exact value of the area of ΔOPQ.

2 marks

Question 16 (3 marks) ⬤⬤⬤

Using vectors, prove that the midpoint of the hypotenuse of a right-angled triangle is equidistant from the 3 vertices of the triangle.

3 marks

Question 17 (4 marks) ⬤⬤⬤

The vectors $8\underset{\sim}{i} + 4\underset{\sim}{j} - \underset{\sim}{k}$ and $4\underset{\sim}{i} - 7\underset{\sim}{j} + 4\underset{\sim}{k}$ are the diagonals of a rectangle.

a   Prove that the rectangle is a square.

2 marks

b   Determine the side length of this square.

2 marks

Question 18 (6 marks) ⬤⬤⬤

Calculate the distance of the point $(-6, 1, 21)$ to the line $\underset{\sim}{r} = \begin{pmatrix} -4 \\ -5 \\ -1 \end{pmatrix} + \lambda \begin{pmatrix} 3 \\ 1 \\ 1 \end{pmatrix}$.

6 marks

Question 19 (6 marks)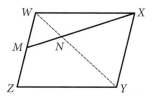

WXYZ is a parallelogram. *M* is the midpoint of *WZ*.

Let $\overrightarrow{WM} = \underset{\sim}{a}$ and $\overrightarrow{WX} = \underset{\sim}{b}$.

a Find $\overrightarrow{WZ}, \overrightarrow{WY}$ and $\overrightarrow{MX}$ in terms of $\underset{\sim}{a}$ and $\underset{\sim}{b}$.   3 marks

b *N* is the point on *MX* such that $MN = \dfrac{1}{3}MX$.   3 marks

Prove that *WNY* is a straight line.

Question 20 (7 marks) ©NESA 2020 HSC EXAM, QUESTION 15(b)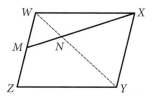

The point *C* divides the interval *AB* so that $\dfrac{CB}{AC} = \dfrac{m}{n}$. The position vectors of *A* and *B* are $\underset{\sim}{a}$ and $\underset{\sim}{b}$ respectively, as shown in the diagram.

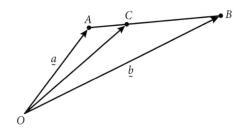

a Show that $\overrightarrow{AC} = \dfrac{n}{m+n}(\underset{\sim}{b} - \underset{\sim}{a})$.   2 marks

b Prove that $\overrightarrow{OC} = \dfrac{m}{m+n}\underset{\sim}{a} + \dfrac{n}{m+n}\underset{\sim}{b}$.   1 mark

Let *OPQR* be a parallelogram with $\overrightarrow{OP} = \underset{\sim}{p}$ and $\overrightarrow{OR} = \underset{\sim}{r}$. The point *S* is the midpoint of *QR* and *T* is the intersection of *PR* and *OS*, as shown in the diagram.

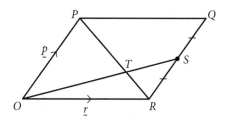

c Show that $\overrightarrow{OT} = \dfrac{2}{3}\underset{\sim}{r} + \dfrac{1}{3}\underset{\sim}{p}$.   3 marks

d Using parts **b** and **c**, or otherwise, prove that *T* is the point that divides the interval   1 mark
PR in the ratio 2:1.

Practice set 1

Worked solutions

1 C

$$d = \sqrt{2^2 + (-3)^2 + 6^2}$$
$$= \sqrt{49}$$
$$= 7$$

2 C

$$d = \sqrt{(4-0)^2 + (-2-0)^2 + (4-0)^2}$$
$$= \sqrt{36}$$
$$= 6$$

3 B

$$x = 2 - \lambda$$

$$y = 5 + 3\lambda$$

Solving simultaneously:

$\lambda = 2 - x$, sub into y:

$$y = 5 + 3(2 - x)$$
$$y = 11 - 3x$$

4 B

Looking for a vector not of the form $\lambda(-2\underset{\sim}{i} + 4\underset{\sim}{j} - 3\underset{\sim}{k})$.

5 C

Vector form $\underset{\sim}{r} = (\underset{\sim}{i} - 3\underset{\sim}{j} + 2\underset{\sim}{k}) + \lambda(2\underset{\sim}{i} - \underset{\sim}{j} + 5\underset{\sim}{k})$.

6 B

Direction vector is
$(1 - -1)\underset{\sim}{i} + (-3 - 2)\underset{\sim}{j} + (2 - -2)\underset{\sim}{k} = 2\underset{\sim}{i} - 5\underset{\sim}{j} + 4\underset{\sim}{k}$
through either of the given points $(1, -3, 2)$
or $(-1, 2, -2)$.

Hence, $(-\underset{\sim}{i} + 2\underset{\sim}{j} - 2\underset{\sim}{k}) + \lambda(2\underset{\sim}{i} - 5\underset{\sim}{j} + 4\underset{\sim}{k})$.

7 D

Checking $(4, -4, 1)$ lies on the line:
$\underset{\sim}{r} = (2\underset{\sim}{i} - \underset{\sim}{j} + 3\underset{\sim}{k}) + \lambda(2\underset{\sim}{i} + 3\underset{\sim}{j} - 4\underset{\sim}{k})$,
$$\therefore 4 = 2 + 2\lambda$$
$$2 = 2\lambda$$

$\lambda = 1$, which is not consistent for
y-, z-components.

8 A

Since scalar product
$2(-1) + 4(3) + (-5)(2) = 0$,
then the lines are perpendicular.

9 B

$$\cos 60° = \frac{2m}{m^2 + 1}$$
$$\frac{1}{2} = \frac{2m}{m^2 + 1}$$
$$m^2 + 1 = 4m$$
$$m^2 - 4m + 1 = 0$$

$$m = \frac{-(-4) \pm \sqrt{(-4)^2 - 4(1)(1)}}{2(1)}$$
$$= \frac{4 \pm \sqrt{12}}{2}$$
$$= \frac{4 \pm 2\sqrt{3}}{2}$$
$$= 2 \pm \sqrt{3}$$

10 C

Since $\underset{\sim}{a} = -\underset{\sim}{c}$, $\underset{\sim}{a} + \underset{\sim}{c} = 0$.

11 D

$$\underset{\sim}{p} = \underset{\sim}{i} + 2\underset{\sim}{j} - \underset{\sim}{k}, \underset{\sim}{q} = m\underset{\sim}{i} - 2\underset{\sim}{j}$$
$$\underset{\sim}{p} - \underset{\sim}{q} = (1 - m)\underset{\sim}{i} + 4\underset{\sim}{j} - \underset{\sim}{k}$$

Hence,
$$1(1 - m) + 2(4) + (-1)(-1) = 0$$
$$10 - m = 0$$
$$m = 10.$$

12 A

$$(2\underset{\sim}{i} - r\underset{\sim}{j} + 5\underset{\sim}{k}) \cdot (3\underset{\sim}{i} - r\underset{\sim}{j} + r\underset{\sim}{k}) = 0$$
$$6 + r^2 + 5r = 0$$
$$(r + 3)(r + 2) = 0$$

$r = -3$ or -2

13 D

$$3n^2 + n - 4 = 0$$
$$(3n + 4)(n - 1) = 0$$

$$n = -\frac{4}{3} \text{ or } 1$$

14 C

For a rhombus, the diagonals are perpendicular.

$$(\underset{\sim}{a} - \underset{\sim}{b}) \cdot (\underset{\sim}{a} + \underset{\sim}{b}) = 0$$

15 B

$$\underset{\sim}{r}(t) = t^2\underset{\sim}{i} - 8\sqrt{t}\,\underset{\sim}{j} + \frac{16}{t}\underset{\sim}{k}$$

$$\dot{\underset{\sim}{r}}(t) = 2t\underset{\sim}{i} - \frac{4}{\sqrt{t}}\underset{\sim}{j} - \frac{16}{t^2}\underset{\sim}{k}$$

$$\therefore \dot{\underset{\sim}{r}}(4) = 2(4)\underset{\sim}{i} - \frac{4}{\sqrt{4}}\underset{\sim}{j} - \frac{16}{4^2}\underset{\sim}{k}$$

So $\dot{\underset{\sim}{r}}(4) = 8\underset{\sim}{i} - 2\underset{\sim}{j} - \underset{\sim}{k}$.

16 C

Since $\underset{\sim}{r} = (5\underset{\sim}{i} + 3\underset{\sim}{k}) + \lambda(-4\underset{\sim}{i} + 3\underset{\sim}{k})$ passes through $(5, 0, 3)$, which is a point on the sphere, we can compute the vector from the centre of the circle:

$$(5, 0, 3) - (2, 0, -1) = (3, 0, 4)$$

Now we check that the line is perpendicular to our vector: $(-4\underset{\sim}{i} + 3\underset{\sim}{k}) \cdot (3\underset{\sim}{i} + 4\underset{\sim}{k}) = 0$.

So this line is a tangent.

17 B

$$\underset{\sim}{r} = (-2\underset{\sim}{i} + 3\underset{\sim}{j}) + 2(a\underset{\sim}{i} + b\underset{\sim}{j}) = (4\underset{\sim}{i} - 5\underset{\sim}{j})$$

Equating components:

$$-2 + 2a = 4$$

$$3 + 2b = -5$$

Hence, $a = 3$, $b = -4$, that is, $\underset{\sim}{v} = (3\underset{\sim}{i} - 4\underset{\sim}{j})$

18 C

Since $\left|\underset{\sim}{A}\right|^2 + \left|\underset{\sim}{B}\right|^2 = \left|\underset{\sim}{C}\right|^2$, then $\underset{\sim}{A}$, $\underset{\sim}{B}$ and $\underset{\sim}{C}$ form a right-angled triangle with hypotenuse $\underset{\sim}{C}$.

---

19 D

Let P and Q be the 2 closest points on r_1 and r_2, respectively.

Hence, $P(1 + 2\lambda, 2 + \lambda, -1 + \lambda)$ and $Q(3 + \mu, -1 + \mu, 2 + \mu)$.

This gives vector $\overrightarrow{PQ} = (2 + \mu - 2\lambda, -3 + \mu - \lambda, 3 + \mu - \lambda)$.

Since $\overrightarrow{PQ} \cdot \underset{\sim}{r_1} = 0$ then $(2 + \mu - 2\lambda)(2) + (-3 + \mu - \lambda)(1) + (3 + \mu - \lambda)(1) = 0$

$$4 + 4\mu - 6\lambda = 0 \quad [1]$$

and $\overrightarrow{PQ} \cdot \underset{\sim}{r_2} = 0$: $\quad (2 + \mu - 2\lambda)(1) + (-3 + \mu - \lambda)(1) + (3 + \mu - \lambda)(1) = 0$

$$2 + 3\mu - 4\lambda = 0 \quad [2]$$

Solving simultaneously:

$2 \times [1]$: $\qquad 8 + 8\mu - 12\lambda = 0 \quad [3]$

$3 \times [2]$: $\qquad 6 + 9\mu - 12\lambda = 0 \quad [4]$

$[4] - [3]$: $\qquad\qquad -2 + \mu = 0$

$$\mu = 2$$

Substitute into $[2]$: $2 + 3(2) - 4\lambda = 0$

$$8 - 4\lambda = 0$$

$$\lambda = 2$$

$\mu = 2$ and $\lambda = 2$

$\therefore P(5, 4, 1)$ and $Q(5, 1, 4)$, giving $\overrightarrow{PQ} = 0\underset{\sim}{i} - 3\underset{\sim}{j} + 3\underset{\sim}{k}$,

$$\left|\overrightarrow{PQ}\right| = \sqrt{0^2 + (-3)^2 + (3)^2}$$

$$= 3\sqrt{2}$$

20 A

Using vector addition $\underset{\sim}{a} + \overrightarrow{AB} = \underset{\sim}{b}$ and $\underset{\sim}{b} + \overrightarrow{BC} = \underset{\sim}{c}$. This gives $\overrightarrow{AB} = \underset{\sim}{b} - \underset{\sim}{a}$ and $\overrightarrow{BC} = \underset{\sim}{c} - \underset{\sim}{b}$.

Given $3\left|\overrightarrow{BC}\right| = \left|\overrightarrow{AB}\right|$, $\quad 3(\underset{\sim}{c} - \underset{\sim}{b}) = \underset{\sim}{b} - \underset{\sim}{a}$,

$$\underset{\sim}{c} = \frac{4}{3}\underset{\sim}{b} - \frac{1}{3}\underset{\sim}{a}.$$

Practice set 2

Worked solutions

Question 1

a Let $\underset{\sim}{u} = 2\underset{\sim}{i} - 3\underset{\sim}{j} + 6\underset{\sim}{k}$,

$$\hat{\underset{\sim}{u}} = \frac{2\underset{\sim}{i} - 3\underset{\sim}{j} + 6\underset{\sim}{k}}{\left|2\underset{\sim}{i} - 3\underset{\sim}{j} + 6\underset{\sim}{k}\right|}$$

$$= \frac{2}{7}\underset{\sim}{i} - \frac{3}{7}\underset{\sim}{j} + \frac{6}{7}\underset{\sim}{k}$$

b $\left|4\underset{\sim}{i} - 2\underset{\sim}{j} + 4\underset{\sim}{k}\right| = \sqrt{4^2 + (-2)^2 + 4^2} = 6$

Question 2

Given $\underset{\sim}{v} = \begin{pmatrix} 4 \\ -2 \\ -4 \end{pmatrix}$, $\quad \hat{\underset{\sim}{v}} = \frac{4}{6}\underset{\sim}{i} - \frac{2}{6}\underset{\sim}{j} - \frac{4}{6}\underset{\sim}{k}$

$$= \frac{2}{3}\underset{\sim}{i} - \frac{1}{3}\underset{\sim}{j} - \frac{2}{3}\underset{\sim}{k}.$$

Question 3

a $\overrightarrow{AB} = \overrightarrow{OB} - \overrightarrow{OA}$

$$= (-\underset{\sim}{i} - 4\underset{\sim}{j} + 5\underset{\sim}{k}) - (2\underset{\sim}{i} - 4\underset{\sim}{j} + 3\underset{\sim}{k})$$

$$= -3\underset{\sim}{i} + 2\underset{\sim}{k}$$

b $\overrightarrow{OB} = \overrightarrow{OA} + \overrightarrow{AB}$

$$= (\underset{\sim}{i} - 4\underset{\sim}{j} - 3\underset{\sim}{k}) + (-2\underset{\sim}{i} + 4\underset{\sim}{j} - \underset{\sim}{k})$$

$$= -\underset{\sim}{i} - 4\underset{\sim}{k}$$

c $\overrightarrow{AO} = \overrightarrow{AB} - \overrightarrow{OB}$

$$= (\underset{\sim}{i} - 3\underset{\sim}{j} + \underset{\sim}{k}) - (3\underset{\sim}{i} + 4\underset{\sim}{j} + \underset{\sim}{k})$$

$$= -2\underset{\sim}{i} - 7\underset{\sim}{j}$$

Question 4

Given $\underset{\sim}{u} = m\underset{\sim}{i} - 4\underset{\sim}{j} + \underset{\sim}{k}$ and $\underset{\sim}{v} = -\underset{\sim}{i} + 2\underset{\sim}{j} + 4\underset{\sim}{k}$,

$\underset{\sim}{u} - \underset{\sim}{v} = (m+1)\underset{\sim}{i} - 6\underset{\sim}{j} - 3\underset{\sim}{k}$

and $\underset{\sim}{u} + \underset{\sim}{v} = (m-1)\underset{\sim}{i} - 2\underset{\sim}{j} + 5\underset{\sim}{k}$.

Since $(\underset{\sim}{u} - \underset{\sim}{v}) \cdot (\underset{\sim}{u} + \underset{\sim}{v}) = 0$, then

$$(m-1)(m+1) + 12 - 15 = 0$$
$$m^2 - 1 - 3 = 0$$
$$m^2 = 4$$
$$\therefore m = \pm 2$$

An alternative approach:

Note that $(\underset{\sim}{u} - \underset{\sim}{v}) \cdot (\underset{\sim}{u} + \underset{\sim}{v}) = \underset{\sim}{u} \cdot \underset{\sim}{u} - \underset{\sim}{v} \cdot \underset{\sim}{v}$,

$$= \left|\underset{\sim}{u}\right|^2 - \left|\underset{\sim}{v}\right|^2$$

so the 2 are perpendicular iff $\left|\underset{\sim}{u}\right| = \left|\underset{\sim}{v}\right|$.

Working through this leads to the same solution.

Question 5

$\underset{\sim}{r} = (2\underset{\sim}{i} - 5\underset{\sim}{j} + \underset{\sim}{k}) + \lambda(-3\underset{\sim}{i} + 4\underset{\sim}{j} + \underset{\sim}{k})$

Question 6

Let vectors be $A = 3\underset{\sim}{i} + 4\underset{\sim}{j} - 2\underset{\sim}{k}$, $B = 7\underset{\sim}{i} + 8\underset{\sim}{j} - 8\underset{\sim}{k}$

and $C = 13\underset{\sim}{i} + 14\underset{\sim}{j} - 17\underset{\sim}{k}$.

For collinearity, $\overrightarrow{AB} = \lambda\overrightarrow{BC}$.

$$\overrightarrow{AB} = \begin{pmatrix} 4 \\ 4 \\ -6 \end{pmatrix} \text{ and } \overrightarrow{BC} = \begin{pmatrix} 6 \\ 6 \\ -9 \end{pmatrix}.$$

From these, we can see that $\overrightarrow{AB} = \frac{2}{3}\overrightarrow{BC}$,

and as they share the point B, the points represented by A, B and C are collinear.

Question 7

$\overrightarrow{PQ} = (-4 - -2)\underset{\sim}{i} + (2-1)\underset{\sim}{j} + (2 - -1)\underset{\sim}{k}$

$$= -2\underset{\sim}{i} + \underset{\sim}{j} + 3\underset{\sim}{k}$$

and $\overrightarrow{PR} = (6 - -2)\underset{\sim}{i} + (-3-1)\underset{\sim}{j} + (-13 - -1)\underset{\sim}{k}$

$$= 8\underset{\sim}{i} - 4\underset{\sim}{j} - 12\underset{\sim}{k}$$

Since $\overrightarrow{PQ} = \lambda\overrightarrow{PR}$, $(-2\underset{\sim}{i} + \underset{\sim}{j} + 3\underset{\sim}{k}) = \lambda(8\underset{\sim}{i} - 4\underset{\sim}{j} - 12\underset{\sim}{k})$,

so we get $\lambda = -\frac{1}{4}$.

Question 8

$$\cos\theta = \frac{(2\underset{\sim}{i} + \underset{\sim}{j} - 2\underset{\sim}{k}) \cdot (3\underset{\sim}{i} - 2\underset{\sim}{j} - \underset{\sim}{k})}{\sqrt{9}\ \sqrt{14}}$$

$$= \frac{6 - 2 + 2}{3\sqrt{14}}$$

$$= \frac{2}{\sqrt{14}}$$

$\theta = 57.6884\ldots°$

$\quad \approx 58°$

Question 9

Let $v = a\underline{i} + b\underline{j} + c\underline{k}$, and given $\underline{u} = 3\underline{i} - 4\underline{j} + 5\underline{k}$.

Since $\underline{u} \cdot \underline{v} = 0$:
$3a - 4b + 5c = 0$.

Choosing $a = n$, $4b = 3n + 5c$.
Letting $c = n$ gives $b = 2n$.

Hence, any vector of the form $n\underline{i} + 2n\underline{j} + n\underline{k}$ (this is simplest case) will be perpendicular to $3\underline{i} - 4\underline{j} + 5\underline{k}$. Other answers are possible, if a different value of c is chosen.

Question 10

To be perpendicular, the direction vectors have a scalar product of 0.

$$(-\underline{i} + 2\underline{j} + 4\underline{k}) \cdot (-m\underline{i} + 2\underline{j} - 4\underline{k}) = 0$$
$$m + 4 - 16 = 0$$
$$m - 12 = 0$$
$$m = 12$$

Question 11

Using γ for the first line, μ for the second, and equating components, we obtain:

$$-5 + 3\gamma = \mu,$$
$$2 + 2\gamma = -6 - 5\mu$$
$$-7 + 6\gamma = -3 - \mu$$

Solving simultaneously:

$$\gamma = 1, \mu = -2$$

Checking for consistency, we get the point of intersection $(-2, 4, -1)$.

Question 12

From each line:

$$2 + 6\gamma_1 = -1 + 6\gamma_2$$
$$1 - 8\gamma_1 = -10 + 7\gamma_2$$
$$-2 + 10\gamma_1 = 5 - 2\gamma_2$$

Solving simultaneously:

$$\gamma_1 = \frac{1}{2}, \gamma_2 = 1$$

Checking for consistency, we get the point of intersection $(5, -3, 3)$.

Question 13

Completing the squares:

$$x^2 - 2x + (1) + y^2 + 4y + (4) + z^2 - 11 = 0 + (5)$$

This gives: $(x - 1)^2 + (y + 2)^2 + (z - 0)^2 = 16$

Therefore, the centre is $C(1, -2, 0)$ and radius $r = 4$.

Question 14

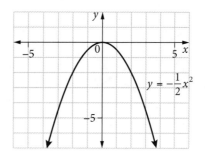

$\underline{r} = t\underline{i} - \frac{1}{2}t^2\underline{j}$, $x = t$, $y = -\frac{1}{2}t^2$

Substituting for t, $y = -\frac{1}{2}x^2$.

Question 15

a $\overrightarrow{PQ} = \overrightarrow{OQ} - \overrightarrow{OP}$

$$= \begin{pmatrix} 0 - -2 \\ 3 - 2 \\ 4 - 1 \end{pmatrix}$$

$$= \begin{pmatrix} 2 \\ 1 \\ 3 \end{pmatrix}$$

$$= 2\underline{i} + \underline{j} + 3\underline{k}$$

b $\cos\theta = \dfrac{(\overrightarrow{OP}) \cdot (\overrightarrow{OQ})}{|\overrightarrow{OP}||\overrightarrow{OQ}|} = \dfrac{0 + 6 + 4}{\sqrt{9}\sqrt{25}} = \dfrac{10}{15} = \dfrac{2}{3}$,

as required.

c Area $\triangle OPQ = \dfrac{1}{2}|\overrightarrow{OP}||\overrightarrow{OQ}|\sin\theta$

$$\sin\theta = \sqrt{1 - \cos^2\theta}$$
$$= \sqrt{1 - \left(\frac{2}{3}\right)^2}$$
$$= \sqrt{1 - \frac{4}{9}}$$
$$= \sqrt{\frac{5}{9}}$$
$$= \frac{\sqrt{5}}{3}$$

So Area $\triangle OPQ = \dfrac{1}{2} \times 3 \times 5 \times \dfrac{\sqrt{5}}{3}$

$$= \frac{5\sqrt{5}}{2} \text{ units}^2.$$

Question 16

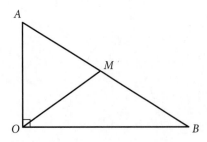

Need to show that $AM = BM = OM$.

$\overrightarrow{OB} = \overrightarrow{OM} + \overrightarrow{MB}$

$\overrightarrow{OA} = \overrightarrow{OM} + \overrightarrow{MA} = \overrightarrow{OM} + \overrightarrow{BM}$

Since $\angle BOA = 90°$, $\overrightarrow{OB} \perp \overrightarrow{OA}$.

Using scalar product $\overrightarrow{OB} \cdot \overrightarrow{OA} = 0$,

that is, $(\overrightarrow{OM} + \overrightarrow{MB}) \cdot (\overrightarrow{OM} + \overrightarrow{BM}) = 0$

$\qquad (\overrightarrow{OM} - \overrightarrow{BM}) \cdot (\overrightarrow{OM} + \overrightarrow{BM}) = 0$.

$OM^2 - BM^2 = 0$ or $OM^2 = BM^2$

$\therefore OM = BM$

Also given that $BM = AM$.

Thus, $OM = BM = AM$, as required.

Question 17

a   The diagonals are $\underset{\sim}{u} - \underset{\sim}{v} = 8\underset{\sim}{i} + 4\underset{\sim}{j} - \underset{\sim}{k}$
and $\underset{\sim}{u} + \underset{\sim}{v} = 4\underset{\sim}{i} - 7\underset{\sim}{j} + 4\underset{\sim}{k}$.

Since $(\underset{\sim}{u} - \underset{\sim}{v}) \cdot (\underset{\sim}{u} + \underset{\sim}{v}) = 32 - 28 - 4 = 0$,

the diagonals are perpendicular
$\Rightarrow$ either a kite or square.

The length of the diagonals:

$\left| \underset{\sim}{u} - \underset{\sim}{v} \right| = \sqrt{8^2 + 4^2 + (-1)^2} = 9$ and

$\left| \underset{\sim}{u} + \underset{\sim}{v} \right| = \sqrt{4^2 + (-7)^2 + (4)^2} = 9$

are equal so the shape is in fact a square.

b   Let the side of the square be length L,

$L^2 + L^2 = 9^2$

$\qquad L^2 = \dfrac{81}{2}$

$\qquad L = \dfrac{9}{\sqrt{2}}$ or $\dfrac{9\sqrt{2}}{2}$.

Question 18

The shortest distance of point $P(-6, 1, 21)$ to the

line $\underset{\sim}{r} = \begin{pmatrix} -4 \\ -5 \\ -1 \end{pmatrix} + \lambda \begin{pmatrix} 3 \\ 1 \\ 1 \end{pmatrix}$ is the perpendicular distance

from P to the line. Let P' be the closest point on
the line to P.

$\overrightarrow{PP'} = \begin{pmatrix} (3\lambda - 4) - (-6) \\ (\lambda - 5) - (1) \\ (\lambda - 1) - (21) \end{pmatrix} = \begin{pmatrix} 3\lambda + 2 \\ \lambda - 6 \\ \lambda - 22 \end{pmatrix}$

Since the shortest distance is the perpendicular
distance:

$\overrightarrow{PP'} \cdot \underset{\sim}{r} = 0, \begin{pmatrix} 3\lambda + 2 \\ \lambda - 6 \\ \lambda - 22 \end{pmatrix} \cdot \begin{pmatrix} 3 \\ 1 \\ 1 \end{pmatrix} = 11\lambda - 22 = 0$

$\therefore \lambda = 2$, hence $\overrightarrow{PP'} = \begin{pmatrix} 3\lambda + 2 \\ \lambda - 6 \\ \lambda - 22 \end{pmatrix} = \begin{pmatrix} 8 \\ -4 \\ -20 \end{pmatrix}$.

Distance of P from P' is

$\left| \overrightarrow{PP'} \right| = \sqrt{8^2 + (-4)^2 + (-20)^2} = \sqrt{480} = 4\sqrt{30}$.

Question 19

a $\overrightarrow{WZ} = 2\underset{\sim}{a}, \overrightarrow{WY} = 2\underset{\sim}{a} + \underset{\sim}{b}, \overrightarrow{MX} = \underset{\sim}{b} - \underset{\sim}{a}$

b   Need to show $\overrightarrow{WN}$ is parallel to $\overrightarrow{WY}$,

that is, $\overrightarrow{WY} = k\overrightarrow{WN}$.

Now, $\overrightarrow{WY} = 2\underset{\sim}{a} + \underset{\sim}{b}$,

$\overrightarrow{WN} = \overrightarrow{WM} + \overrightarrow{MN}$

$\qquad = \underset{\sim}{a} + \dfrac{1}{3}(\overrightarrow{MX})$

$\qquad = \underset{\sim}{a} + \dfrac{1}{3}(\underset{\sim}{b} - \underset{\sim}{a})$

$\qquad = \dfrac{1}{3}(2\underset{\sim}{a} + \underset{\sim}{b})$

$\qquad = \dfrac{1}{3}(\overrightarrow{WY})$, as required.

Question 20

a $\dfrac{CB}{AC} = \dfrac{m}{n}$

So AB is divided into $m + n$ parts.

Then

$$\overrightarrow{AB} = \overrightarrow{AC} + \overrightarrow{CB}$$

$$= \left(\dfrac{n}{m+n}\right)\overrightarrow{AB} + \left(\dfrac{m}{m+n}\right)\overrightarrow{AB}.$$

Now,

$$\underset{\sim}{a} + \overrightarrow{AB} = \underset{\sim}{b}$$

$$\overrightarrow{AB} = \underset{\sim}{b} - \underset{\sim}{a}$$

so

$$\overrightarrow{AC} = \overrightarrow{AB}\left(\dfrac{n}{m+n}\right)$$

$$= (\underset{\sim}{b} - \underset{\sim}{a})\left(\dfrac{n}{m+n}\right).$$

b $\overrightarrow{OC} = \underset{\sim}{a} + \overrightarrow{AC}$

$$= \dfrac{\underset{\sim}{a}(m+n)}{m+n} + \left(\dfrac{n}{m+n}\right)\underset{\sim}{b} - \left(\dfrac{n}{m+n}\right)\underset{\sim}{a}$$

$$= \dfrac{\underset{\sim}{a}m + \underset{\sim}{a}n + \underset{\sim}{b}n - \underset{\sim}{a}n}{m+n}$$

$$= \dfrac{\underset{\sim}{a}m + \underset{\sim}{b}n}{m+n}$$

$$= \left(\dfrac{m}{m+n}\right)\underset{\sim}{a} + \left(\dfrac{n}{m+n}\right)\underset{\sim}{b}$$

c T is the point of intersection of OS and PR.

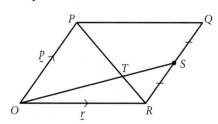

Equation of OS is $\lambda_1\left(\underset{\sim}{r} + \dfrac{1}{2}\underset{\sim}{p}\right)$ [1]

because $\underset{\sim}{p} = \overrightarrow{RQ}$ (opposite sides of a parallelogram).

Equation of PR is $\underset{\sim}{r} + \lambda_2(\underset{\sim}{p} - \underset{\sim}{r})$ [2]

[or $\underset{\sim}{p} + \lambda_2(\underset{\sim}{r} - \underset{\sim}{p})$].

Equating [1] and [2]:

$\underset{\sim}{r}:$ $\lambda_1 = 1 - \lambda_2$ [3]

$\underset{\sim}{p}:$ $\dfrac{1}{2}\lambda_1 = \lambda_2$ [4]

Substitute [4] into [3]:

$$\lambda_1 = 1 - \dfrac{1}{2}\lambda_1$$

$$\dfrac{3}{2}\lambda_1 = 1$$

$$\lambda_1 = \dfrac{2}{3}$$

Substitute into [4]:

$$\lambda_2 = \dfrac{1}{2}\left(\dfrac{2}{3}\right) = \dfrac{1}{3}$$

Substitute into [1]:

$$\overrightarrow{OT} = \dfrac{2}{3}\left(\underset{\sim}{r} + \dfrac{1}{2}\underset{\sim}{p}\right)$$

$$= \dfrac{2}{3}\underset{\sim}{r} + \dfrac{1}{3}\underset{\sim}{p}$$

Alternative method using similar triangles:

In $\triangle POT$ and $\triangle RST$:

$\angle POT = \angle RST$ (alternate angles, $PO \parallel QR$)

$\angle PTO = \angle RTS$ (vertically opposite angles)

$\therefore \triangle POT \;|||\; \triangle RST$ (equiangular)

$\dfrac{SR}{PO} = \dfrac{SR}{QR} = \dfrac{1}{2}$ (opposite sides of a parallelogram)

$\therefore \dfrac{SR}{PO} = \dfrac{ST}{OT} = \dfrac{1}{2}$ $\begin{matrix}\text{(matching sides in}\\\text{similar triangles)}\end{matrix}$

$$\overrightarrow{OS} = \overrightarrow{OR} + \overrightarrow{RS}$$

$$= \underset{\sim}{r} + \dfrac{1}{2}\underset{\sim}{p}$$

As $\dfrac{OT}{ST} = \dfrac{2}{1},$

$$\dfrac{OT}{OS} = \dfrac{OT}{OT + TS} = \dfrac{2}{2+1} = \dfrac{2}{3}.$$

$$\overrightarrow{OT} = \dfrac{2}{3}\overrightarrow{OS}$$

$$= \dfrac{2}{3}\left(\underset{\sim}{r} + \dfrac{1}{2}\underset{\sim}{p}\right)$$

$$= \dfrac{2}{3}\underset{\sim}{r} + \dfrac{1}{3}\underset{\sim}{p}$$

d $\overrightarrow{OT} = \frac{2}{3}\underset{\sim}{r} + \frac{1}{3}\underset{\sim}{p}$

This is the same situation as part **b**,

$\overrightarrow{OC} = \frac{m}{m+n}\underset{\sim}{a} + \frac{m}{m+n}\underset{\sim}{b}$, where $m = 2$, $n = 1$.

$\therefore \frac{PT}{TR} = \frac{m}{n} = \frac{2}{1}$

$\therefore$ T divides PR in the ratio $2:1$.

OR

$\overrightarrow{PT} = \overrightarrow{OT} - \underset{\sim}{p}$

$\qquad = \frac{2}{3}\underset{\sim}{r} + \frac{1}{3}\underset{\sim}{p} - \underset{\sim}{p}$ (from part **c**)

$\qquad = \frac{2}{3}\underset{\sim}{r} - \frac{2}{3}\underset{\sim}{p}$

$\overrightarrow{PR} = \underset{\sim}{r} - \underset{\sim}{p}$

$\therefore \overrightarrow{PT} = \frac{2}{3}\overrightarrow{PR}$

$\therefore$ T divides PR in the ratio $2:1$.

Alternative method using similar triangles:

From part **c**:

$\frac{RT}{PT} = \frac{1}{2}$ (matching sides in similar triangles)

$\therefore$ $PT:TR = 2:1$ and T divides PR in the ratio $2:1$.

HSC exam topic grid (2020)

This table shows the coverage of this topic in past HSC exams by question number. The past exams can be downloaded from the NESA website (www.educationstandards.nsw.edu.au) by selecting 'Year 11 – Year 12', 'HSC exam papers'. NESA marking feedback and guidelines can also be found there.

The new Mathematics Extension 2 course was first examined in 2020. For exams before 2020, select 'Year 11 – Year 12', 'Resources archive', 'HSC exam papers archive'.

Vectors were introduced to the Mathematics Extension 1 and 2 courses in 2020.

	Operations with vectors	Geometrical proofs	Vector equations of lines	Vector equations of curves and spheres
2020 new course	1, 11(d)	**15(b)**	3, 11(d), **13(b)**	

Questions in **bold** can be found in this chapter.

CHAPTER 3
COMPLEX NUMBERS

COMPLEX NUMBERS

Complex numbers

- Cartesian form $z = a + ib$
- Re(z), Im(z)
- Adding, subtracting, multiplying, dividing
- Complex conjugate, $\bar{z}$
- Realising the denominator
- Reciprocal
- Square root of a complex number

Euler's formula and exponential form

- Euler's formula $e^{i\theta} = \cos\theta + i\sin\theta$
- Exponential form $z = re^{i\theta}$
- Exponential, Cartesian, polar forms
- Powers of complex numbers in exponential form

Complex numbers as vectors

- Adding and subtracting complex numbers
- Multiplying complex numbers
- Conjugates, multiplying by a scalar, multiplying by i

The complex plane and polar form

- The complex plane, Argand diagram
- Polar (modulus-argument) form
- Properties of modulus and argument
- Multiplying and dividing complex numbers
- Powers of complex numbers

De Moivre's theorem and solving equations

- De Moivre's theorem
- Solving quadratic equations
- Quadratic equations with complex coefficients
- Polynomial equations and conjugate roots

Roots, curves and regions

- Roots of unity: location on the unit circle
- Roots of a complex number
- Graphing lines and curves in the complex plane: equations involving modulus and argument
- Graphing regions in the complex plane: inequalities

Glossary

Argand diagram
A diagram used to represent geometrically the complex number $z = a + ib$ as the point $P(a, b)$ or the vector z or $\overrightarrow{OP}$ on the complex plane.

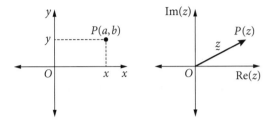

argument
The angle that the vector representing the complex number z makes with the positive x-axis of the complex plane, abbreviated as arg z or θ. For $z = a + ib$, $\tan \theta = \dfrac{b}{a}$.

See also **principal argument.**

ℂ
The set of complex numbers.

Cartesian (or rectangular) form
The notation $z = a + ib$ for a complex number.

complex conjugate
The conjugate of $z = a + ib$ is $\overline{z} = a - ib$.

complex number
A number that can be written in the form $a + ib$, where a and b are real numbers. It is a member of ℂ, the set of complex numbers.

complex plane
A number plane for graphing complex numbers. Also called **Argand diagram.**

De Moivre's theorem
$(\cos \theta + i \sin \theta)^n = \cos n\theta + i \sin n\theta, \forall\, n \in \mathbb{Z}$

A+ DIGITAL FLASHCARDS
Revise this topic's key terms and concepts by scanning the QR code or typing the URL into your browser.

https://get.ga/a-hsc-maths-ext-2

exponential form
The notation $z = re^{i\theta}$ for a complex number, which involves Euler's formula.

Euler's formula
For any real θ, $e^{i\theta} = \cos \theta + i \sin \theta$.

imaginary number, i
The number such that $i = \sqrt{-1}$, which implies $i^2 = -1$.

imaginary part
The imaginary part of $z = a + ib$ is $\text{Im}(z) = b$.

locus
A set of points that obey a certain condition. Its graph can be a line, ray, curve or region.

modulus
The length or magnitude of the vector representing a complex number in the complex plane, abbreviated mod z, $|z|$ or r. For $z = a + ib$, $r = |z| = \sqrt{a^2 + b^2}$.

polar (or modulus-argument) form
The notation $z = r(\cos \theta + i \sin \theta)$ for a complex number.

principal argument
The argument of the complex number z in the interval $(-\pi, \pi]$, a specific angle, abbreviated as Arg z.

real part
The real part of $z = a + ib$ is $\text{Re}(z) = a$.

root
A solution to an equation.

root of unity
A solution to the equation $z^n = 1$, $n \in \mathbb{N}$.

vector
A quantity with both magnitude and direction, represented graphically by an arrow with specific length and direction.

Topic summary

Introduction to complex numbers (MEX-N1)

N1.1 Arithmetic of complex numbers

Imaginary numbers and complex numbers

The **imaginary number** i is the number such that $i = \sqrt{-1}$ or $i^2 = -1$.

A **complex number** z is a number that can be written in the form $a + ib$, where a and b are real numbers.

The **real part** of $z = a + ib$ is denoted by $\mathrm{Re}(z)$, where $\mathrm{Re}(z) = a$.

The **imaginary part** of $z = a + ib$ is denoted by $\mathrm{Im}(z)$, where $\mathrm{Im}(z) = b$.

If $\mathrm{Re}(z) = 0$, then z is **purely imaginary**, for example, $2i$, $-5i$ and $-\sqrt{10}i$.

If $\mathrm{Im}(z) = 0$, then z is **purely real** or just **real**, for example, 12, $\sqrt{7}$ and $2 - \sqrt{3}$.

The set of complex numbers is shown as $\mathbb{C}$, and includes $\mathbb{R}$, the set of real numbers. All real numbers are also complex numbers.

Example 1

Simplify:

a i^2 **b** i^3 **c** i^4 **d** i^{10}

Solution

a $i^2 = -1$ (by definition)

b $\begin{aligned} i^3 &= i^2 \times i \\ &= -1 \times i \\ &= -i \end{aligned}$

c $\begin{aligned} i^4 &= i^2 \times i^2 \\ &= (-1) \times (-1) \\ &= 1 \end{aligned}$

d $\begin{aligned} i^{10} &= i^4 \times i^4 \times i^2 \\ &= 1 \times 1 \times (-1) \quad \text{from } \mathbf{c} \\ &= -1 \end{aligned}$ OR $\begin{aligned} i^{10} &= (i^2)^5 \\ &= (-1)^5 \\ &= -1 \end{aligned}$

Equivalence of complex numbers

For 2 complex numbers $a + ib$ and $c + id$ (where a, b, c and d are real numbers), $a + ib = c + id$ if and only if $a = c$ and $b = d$.

Example 2

Find the values of x and y such that $2x - 6i - 2yi + 10 = 0$, where x and y are real.

Solution

$2x - 6i - 2yi + 10 = 0$ means $2x - 6i - 2yi + 10 = 0 + 0i$

$$(2x + 10) + i(-6 - 2y) = 0 + 0i$$

Equating real and imaginary parts:

$$\therefore \; 2x + 10 = 0 \qquad\qquad -6 - 2y = 0$$
$$2x = -10 \qquad\qquad -6 = 2y$$
$$x = -5 \qquad\qquad y = -3$$

TOPIC SUMMARY

Complex conjugate

For $z = a + ib$, its **complex conjugate** is denoted by $\bar{z}$ and $\bar{z} = a - ib$.

The product $z\bar{z}$ is real.

$$(-1 + 3i)(-1 - 3i) = (-1)^2 - (3i)^2$$
$$= 1 - 9i^2$$
$$= 1 - 9(-1)$$
$$= 1 + 9$$
$$= 10$$

Realising the denominator

If a complex number z has a complex denominator, we can make the denominator a real number by multiplying z by $\dfrac{\bar{z}}{\bar{z}}$. This is called **realising the denominator**.

Example 3

Simplify $\dfrac{2 + 2i\sqrt{3}}{\sqrt{3} - i}$ by realising the denominator.

Solution

$$\frac{2 + 2i\sqrt{3}}{\sqrt{3} - i} = \frac{2 + 2i\sqrt{3}}{\sqrt{3} - i} \times \frac{\sqrt{3} + i}{\sqrt{3} + i}$$
$$= \frac{2\sqrt{3} + 2i + 6i + 2i^2\sqrt{3}}{3 - i^2}$$
$$= \frac{2\sqrt{3} + 8i - 2\sqrt{3}}{3 + 1}$$
$$= \frac{8i}{4}$$
$$= 2i$$

Square root of a complex number

To find $\sqrt{a + ib}$, let $a + ib = (x + iy)^2$ $(x, y \in \mathbb{R})$, then equate real and imaginary parts.

Example 4

Find $\sqrt{5 + 12i}$.

Solution

Let $\sqrt{5 + 12i} = x + iy$, where x and y are real.

Then squaring both sides,

$$5 + 12i = (x + iy)^2$$
$$= x^2 + 2xyi + i^2y^2$$
$$= x^2 - y^2 + 2xyi$$

Equating real and imaginary parts,

$$5 = x^2 - y^2$$
$$12 = 2xy$$
$$xy = 6$$

Solving simultaneously:

$$x^2 - y^2 = 5 \quad [1]$$
$$xy = 6 \quad [2]$$

From [2]:

$$y = \frac{6}{x} \quad [3]$$

> **Hint**
> Where possible, such simultaneous equations should be quickly solved by inspection. $x = 3, y = 2$, or $x = -3, y = -2$.

Substitute into [1]:

$$x^2 - \left(\frac{6}{x}\right)^2 = 5$$
$$x^2 - \frac{36}{x^2} = 5$$
$$x^4 - 36 = 5x^2$$
$$x^4 - 5x^2 - 36 = 0$$
$$(x^2 - 9)(x^2 + 4) = 0$$
$$x^2 - 9 = 0 \qquad x \in \mathbb{R}$$
$$x = \pm 3$$

Substitute into [3] to find y:

When $x = 3$, $y = \dfrac{6}{3} = 2$.

When $x = -3$, $y = \dfrac{6}{-3} = -2$.

Therefore, $\sqrt{5 + 12i} = 3 + 2i$ or $-3 - 2i$.

Notice that the 2 square roots are the negative of each other.

N1.2 Geometric representation of a complex number

The Argand diagram and complex plane

A complex number $z = a + ib$ can be represented geometrically on a number plane called an **Argand diagram** or **complex plane**, with a horizontal axis denoted by x or $\text{Re}(z)$ and a vertical axis denoted by y or $\text{Im}(z)$.

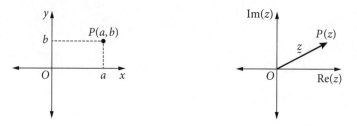

z is represented by the point $P(a, b)$ or the vector $\overrightarrow{OP}$.

A **vector** is a quantity with both magnitude and direction, and is represented graphically by an arrow with a specific length and direction.

The modulus and argument

The **modulus** of a complex number is the length of the corresponding vector on the complex plane.

The modulus, r, of $z = a + ib$ is written as **mod** z or $|z|$.

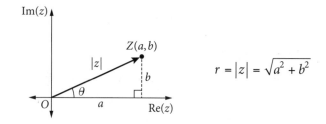

$$r = |z| = \sqrt{a^2 + b^2}$$

The **argument** of a complex number is the angle the vector makes with the positive x-axis.

The argument, θ, of $z = a + ib$ is written as **arg** z.

$$\tan \theta = \frac{b}{a}, \text{ where } \theta = \arg z$$

If $z = 0 + 0i$, then $\arg z$ is undefined.

Arg z is the **principal argument** of z in the interval $(-\pi, \pi]$.

The polar form (modulus-argument form) of a complex number

$$z = r(\cos \theta + i \sin \theta)$$

When z is expressed in terms of x and y, that is, $x + iy$, this is called the **Cartesian form** or **rectangular form** because it is based on the number plane.

9780170459266

Example 5

Express each complex number in polar form.

a $z = -\sqrt{2} + i\sqrt{2}$

b $z = 2\left(\cos\dfrac{\pi}{6} - i\sin\dfrac{\pi}{6}\right)$

Solution

a $a = -\sqrt{2}, b = \sqrt{2}$

$r = \sqrt{\left(-\sqrt{2}\right)^2 + \left(\sqrt{2}\right)^2}$

$\quad = \sqrt{4}$

$\quad = 2$

$\tan\theta = \dfrac{\sqrt{2}}{-\sqrt{2}} = -1$ and $z = -\sqrt{2} + i\sqrt{2}$ lies in

the 2nd quadrant, so $\theta = \dfrac{3\pi}{4}$ (in the interval $-\pi, \pi]$).

In polar form, $z = 2\left(\cos\dfrac{3\pi}{4} + i\sin\dfrac{3\pi}{4}\right)$.

b $z = 2\left(\cos\dfrac{\pi}{6} - i\sin\dfrac{\pi}{6}\right)$ is not in polar form as it

is not in the form $z = r(\cos\theta + i\sin\theta)$ where
a '+' separates the cos and sin terms. However,
use $\cos(-\theta) = \cos\theta$ and $\sin(-\theta) = -\sin\theta$ to
express z in polar form.

$z = 2\left(\cos\dfrac{\pi}{6} - i\sin\dfrac{\pi}{6}\right)$

$\quad = 2\left[\cos\left(-\dfrac{\pi}{6}\right) + i\sin\left(-\dfrac{\pi}{6}\right)\right]$

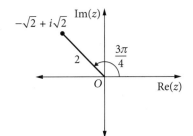

The conjugate in polar form

The conjugate of $z = r(\cos\theta + i\sin\theta)$ is $\overline{z} = r(\cos\theta - i\sin\theta)$, which can be written in polar form as

$$\overline{z} = r\left[\cos(-\theta) + i\sin(-\theta)\right].$$

Example 6

Express $z = 3\left(\cos\dfrac{5\pi}{3} + i\sin\dfrac{5\pi}{3}\right)$ in Cartesian form.

Solution

To convert from polar form to Cartesian form, simply evaluate and expand.

$$z = 3\left(\cos\dfrac{5\pi}{3} + i\sin\dfrac{5\pi}{3}\right)$$

$$= 3\left[\dfrac{1}{2} + i\left(-\dfrac{\sqrt{3}}{2}\right)\right]$$

$$= \dfrac{3}{2} - \dfrac{3\sqrt{3}i}{2}$$

Properties of moduli and arguments

1. Product of 2 complex numbers

   Let $z_1 = r_1(\cos\theta_1 + i\sin\theta_1)$ and $z_2 = r_2(\cos\theta_2 + i\sin\theta_2)$.

   $$z_1 z_2 = r_1 r_2 [\cos(\theta_1 + \theta_2) + i\sin(\theta_1 + \theta_2)]$$

2. Quotient of 2 complex numbers

   $$\frac{z_1}{z_2} = \frac{r_1}{r_2}\left[\cos(\theta_1 - \theta_2) + i\sin(\theta_1 - \theta_2)\right], z_2 \neq 0$$

3. Power of a complex number

   Let $z = r(\cos\theta + i\sin\theta)$.

   $$z^n = r^n(\cos n\theta + i\sin n\theta), \text{ where } n \text{ is an integer.}$$

4. Reciprocal of a complex number

   $$z^{-1} = r^{-1}(\cos\theta - i\sin\theta), z \neq 0$$

5. Negative power of a complex number

   $$z^{-n} = r^{-n}[\cos(-n\theta) + i\sin(-n\theta)], z \neq 0, \text{ where } n \text{ is an integer.}$$

6. Product of complex numbers

   Let $z_1 = r_1(\cos\theta_1 + i\sin\theta_1)$, $z_2 = r_2(\cos\theta_2 + i\sin\theta_2)$, $\cdots$, $z_n = r_n(\cos\theta_n + i\sin\theta_n)$.

   $$z_1 z_2 z_3 \cdots z_n = r_1 r_2 r_3 \cdots r_n[\cos(\theta_1 + \theta_2 + \theta_3 + \cdots + \theta_n) + i\sin(\theta_1 + \theta_2 + \theta_3 + \cdots + \theta_n)],$$
   where n is an integer.

The above properties can also be written separately in terms of their moduli and arguments:

1. $\left|z_1\right|\left|z_2\right| = \left|z_1 z_2\right|$ and $\arg z_1 z_2 = \arg z_1 + \arg z_2$

2. $\dfrac{\left|z_1\right|}{\left|z_2\right|} = \left|\dfrac{z_1}{z_2}\right|$ and $\arg\dfrac{z_1}{z_2} = \arg z_1 - \arg z_2$

3. $\left|z\right|^n = \left|z^n\right|$ and $\arg z^n = n \times \arg(z)$ for $n \in \mathbb{Z}$

4. $\left|z^{-1}\right| = \dfrac{1}{\left|z\right|}$ and $\arg z^{-1} = -\arg z, z \neq 0$

5. $\left|z^{-n}\right| = \dfrac{1}{\left|z\right|^n}$ and $\arg z^{-n} = -n\arg z, z \neq 0$ for $n \in \mathbb{Z}$

6. $\left|z_1\right|\left|z_2\right|\left|z_3\right|\cdots\left|z_n\right| = \left|z_1 z_2 z_3 \cdots z_n\right|$ and $\arg z_1 z_2 \cdots z_n = \arg z_1 + \arg z_2 + \cdots + \arg z_n$, where $n \in \mathbb{N}$.

Example 7

For $z_1 = 2\left(\cos\dfrac{\pi}{5} + i\sin\dfrac{\pi}{5}\right)$ and $z_2 = 5\left(\cos\dfrac{\pi}{7} + i\sin\dfrac{\pi}{7}\right)$, evaluate:

a $z_1 z_2$ **b** $\dfrac{z_1}{z_2}$ **c** $\dfrac{1}{z_1}$ **d** $(z_2)^7$

Solution

a Using $z_1 z_2 = r_1 r_2[\cos(\theta_1 + \theta_2) + i\sin(\theta_1 + \theta_2)]$

$$z_1 z_2 = 2\left(\cos\frac{\pi}{5} + i\sin\frac{\pi}{5}\right) \times 5\left(\cos\frac{\pi}{7} + i\sin\frac{\pi}{7}\right)$$

$$= (2 \times 5)\left[\cos\left(\frac{\pi}{5} + \frac{\pi}{7}\right) + i\sin\left(\frac{\pi}{5} + \frac{\pi}{7}\right)\right]$$

$$= 10\left[\cos\frac{12\pi}{35} + i\sin\frac{12\pi}{35}\right]$$

b Using $\dfrac{z_1}{z_2} = \dfrac{r_1}{r_2}\left[\cos(\theta_1 - \theta_2) + i\sin(\theta_1 - \theta_2)\right]$

$$\frac{z_1}{z_2} = \frac{2\left(\cos\dfrac{\pi}{5} + i\sin\dfrac{\pi}{5}\right)}{5\left(\cos\dfrac{\pi}{7} + i\sin\dfrac{\pi}{7}\right)}$$

$$= \frac{2}{5}\left[\cos\left(\frac{\pi}{5} - \frac{\pi}{7}\right) + i\sin\left(\frac{\pi}{5} - \frac{\pi}{7}\right)\right]$$

$$= \frac{2}{5}\left[\cos\frac{2\pi}{35} + i\sin\frac{2\pi}{35}\right]$$

c Using $z^{-1} = r^{-1}(\cos\theta - i\sin\theta)$

$$\frac{1}{z_1} = \frac{1}{2\left(\cos\dfrac{\pi}{5} + i\sin\dfrac{\pi}{5}\right)}$$

$$= \frac{1}{2}\left(\cos\frac{\pi}{5} - i\sin\frac{\pi}{5}\right)$$

d Using $z^n = r^n(\cos n\theta + i\sin n\theta)$

$$(z_2)^7 = 5^7\left(\cos\frac{7\pi}{7} + i\sin\frac{7\pi}{7}\right)$$

$$= 5^7\left(\cos\pi + i\sin\pi\right)$$

$$= 5^7\left[-1 + i(0)\right]$$

$$= -5^7$$

Properties of conjugates

1. Reciprocal of a complex number with modulus 1

   If $z = \cos\theta + i\sin\theta$, then

   $$z^{-1} = \overline{z}.$$

2. Product of complex conjugate pairs

   For any complex number z,

   $$z\overline{z} = |z|^2 \text{ and } \arg z\overline{z} = 0.$$

3. Modulus and argument of a complex conjugate

   $$|\overline{z}| = |z| \text{ and } \arg\overline{z} = -\arg z$$

4. Sum of conjugates of 2 complex numbers

   $$\overline{z_1} + \overline{z_2} = \overline{z_1 + z_2}$$

5. Product of conjugates of 2 complex numbers

   $$\overline{z_1}\,\overline{z_2} = \overline{z_1 z_2}$$

6. Sum of conjugate pairs

   $$z + \overline{z} = 2\,\mathrm{Re}(z)$$

7. Difference of conjugate pairs

   $$z - \overline{z} = 2i\,\mathrm{Im}(z)$$

The triangle inequality

For 2 complex numbers z and w,

$$|z + w| \le |z| + |w|.$$

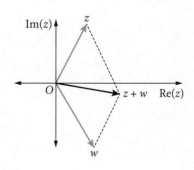

N1.3 Other representations of complex numbers

Euler's formula

For any real θ,

$$e^{i\theta} = \cos\theta + i\sin\theta.$$

This is called the **exponential form** of a complex number.

Example 8

Use Euler's formula to simplify $\left(\cos\dfrac{\pi}{4} + i\sin\dfrac{\pi}{4}\right)\left(\cos\dfrac{2\pi}{3} + i\sin\dfrac{2\pi}{3}\right)$.

Solution

$$\left(\cos\frac{\pi}{4} + i\sin\frac{\pi}{4}\right)\left(\cos\frac{2\pi}{3} + i\sin\frac{2\pi}{3}\right)$$

$$= e^{\frac{i\pi}{4}} \times e^{\frac{2i\pi}{3}}$$

$$= e^{\frac{i\pi}{4} + \frac{2i\pi}{3}}$$

$$= e^{\frac{11i\pi}{12}}$$

$$= \cos\frac{11\pi}{12} + i\sin\frac{11\pi}{12}$$

Example 9

If $z = re^{i\theta}$, prove that $\arg z^n = n\arg z$.

Solution

$$z = re^{i\theta}$$

$$z^n = (re^{i\theta})^n = r^n e^{in\theta}$$

$$\arg z^n = \arg(r^n e^{in\theta})$$

$$= n\theta$$

$$= n\arg z \qquad [\theta = \arg z]$$

Using complex numbers (MEX-N2)

N2.1 Solving equations with complex numbers

De Moivre's theorem

For $z = (\cos\theta + i\sin\theta) = e^{i\theta}$:

$$(\cos\theta + i\sin\theta)^n = \cos n\theta + i\sin n\theta = e^{in\theta} \quad \text{(polar form)}$$
$$(e^{i\theta})^n = e^{in\theta} \quad \text{(exponential form)}$$

For example,

$$\left[2(\cos\theta + i\sin\theta)\right]^8 = 2^8(\cos 8\theta + i\sin 8\theta)$$
$$= 256(\cos 8\theta + i\sin 8\theta)$$

$$\left[8(\cos\pi + i\sin\pi)\right]^{\frac{1}{3}} = 8^{\frac{1}{3}}\left(\cos\frac{\pi}{3} + i\sin\frac{\pi}{3}\right)$$
$$= 2\left(\cos\frac{\pi}{3} + i\sin\frac{\pi}{3}\right)$$

Example 10

Use De Moivre's theorem and the binomial expansion of $(a + b)^3$ to prove that:

$$\sin 3\theta = 3\sin\theta - 4\sin^3\theta.$$

Solution

Use De Moivre's theorem:

$$(\cos\theta + i\sin\theta)^3 = \cos 3\theta + i\sin 3\theta$$

Using the binomial expansion

$$(a + b)^3 = a^3 + 3a^2b + 3ab^2 + b^3,$$

we have:

$$(\cos\theta + i\sin\theta)^3 = \cos^3\theta + 3i\cos^2\theta\sin\theta - 3\cos\theta\sin^2\theta - i\sin^3\theta$$

Equating both expansions:

$$\therefore \cos 3\theta + i\sin 3\theta = \cos^3\theta + 3i\cos^2\theta\sin\theta - 3\cos\theta\sin^2\theta - i\sin^3\theta$$

Equating imaginary parts we have:

$$\sin 3\theta = 3\cos^2\theta\sin\theta - \sin^3\theta$$
$$= 3(1 - \sin^2\theta)\sin\theta - \sin^3\theta$$
$$= 3\sin\theta - 3\sin^3\theta - \sin^3\theta \qquad \text{in terms of } \sin\theta \text{ only}$$
$$= 3\sin\theta - 4\sin^3\theta, \text{ as required.}$$

The identities $z^n + \dfrac{1}{z^n}$ and $z^n - \dfrac{1}{z^n}$

If $z = \cos\theta + i\sin\theta$, then $\forall n \in \mathbb{N}$:

$$z^n + \frac{1}{z^n} = 2\cos n\theta$$

$$z^n - \frac{1}{z^n} = 2i\sin n\theta$$

Quadratic equations

Example 11

Solve $x^2 + 2x + 10 = 0$ using the quadratic formula.

Solution

$$x^2 + 2x + 10 = 0$$

$$x = \frac{-b \pm \sqrt{b^2 - 4ac}}{2a}$$

$$x = \frac{-2 \pm \sqrt{2^2 - 4(1)(10)}}{2(1)}$$

$$= \frac{-2 \pm \sqrt{-36}}{2}$$

$$= \frac{-2 \pm 6i}{2}$$

$$= -1 \pm 3i$$

> **Hint**
> Note that the roots are **complex conjugates**.

Quadratic equations with complex coefficients

Example 12

Solve $x^2 - (5 - 2i)x + 5 - 5i = 0$.

Solution

Using the quadratic formula where $a = 1$, $b = -(5 - 2i)$ and $c = 5 - 5i$:

$$x = \frac{(5 - 2i) \pm \sqrt{(5 - 2i)^2 - 4(1)(5 - 5i)}}{2(1)}$$

$$= \frac{(5 - 2i) \pm \sqrt{25 - 20i - 4 - 20 + 20i}}{2}$$

$$= \frac{(5 - 2i) \pm \sqrt{1}}{2}$$

$$= \frac{5 - 2i \pm 1}{2}$$

$$= \frac{6 - 2i}{2} \text{ or } \frac{4 - 2i}{2}$$

$$= 3 - i \text{ or } 2 - i$$

Polynomial equations

Complex conjugate root theorem

If a polynomial equation $P(z) = 0$ has real coefficients and if $\alpha = a + ib$ (where $a, b \in \mathbb{R}$) is a root, then $\bar{\alpha} = a - ib$ is also a root of $P(z) = 0$.

Real and complex roots

Given a polynomial equation $P(z) = 0$ with real coefficients and of degree n:

1. if n is odd, then $P(z) = 0$ has at least one real root and the complex roots will come in conjugate pairs

2. if n is even, then $P(z) = 0$ has an even number of real roots or no real roots, and the complex roots will come in conjugate pairs

3. if $P(z) = 0$ has complex roots $\alpha = a + ib$ and $\bar{\alpha} = a - ib$, then $P(z)$ will have a quadratic factor of the form $(x - \alpha)(x - \bar{\alpha}) = \left[x^2 - (\alpha + \bar{\alpha})x + \alpha\bar{\alpha} \right]$.

TOPIC SUMMARY

Example 13

Consider the polynomial $P(x) = x^3 - 4x^2 + 6x - 4$.

a Show that $x = 2$ is a root of $x^3 - 4x^2 + 6x - 4 = 0$.

b Hence, solve the equation $x^3 - 4x^2 + 6x - 4 = 0$.

Solution

a Using the factor theorem, show that $P(2) = 0$.

$$P(2) = 2^3 - 4(2^2) + 6(2) - 4$$
$$= 8 - 16 + 12 - 4$$
$$= 0$$

So $x = 2$ is a root.

b To solve the equation we factorise $P(x)$ further as we now know $(x - 2)$ is a factor.

We could use polynomial division by $(x - 2)$ but an easier way is to use the fact that:

$$x^3 - 4x^2 + 6x - 4 = (x - 2)(x^2 + bx + c)$$

and equate coefficients to solve for b and c.

$$x^3 - 4x^2 + 6x - 4 = x^3 + bx^2 + cx - 2x^2 - 2bx - 2c$$
$$= x^3 + (b - 2)x^2 + (c - 2b)x - 2c$$

$b - 2 = -4 \qquad -2c = -4$
$\qquad b = -2 \qquad\quad c = 2 \qquad \therefore x^3 - 4x^2 + 6x - 4 = (x - 2)(x^2 - 2x + 2)$

We can now factorise fully to find all of the roots.

By completing the square (or using the quadratic formula):

$$(x - 2)(x^2 - 2x + 2) = (x - 2) \left[(x^2 - 2x + 1) + 1 \right]$$
$$= (x - 2)[(x - 1)^2 + 1]$$
$$\therefore x^3 - 4x^2 + 6x - 4 = (x - 2)[(x - 1)^2 + 1] = 0$$

$x - 2 = 0 \qquad \text{or} \qquad (x - 1)^2 + 1 = 0$
$\qquad x = 2 \qquad \text{or} \qquad (x - 1)^2 = -1$
$\qquad\qquad\qquad\qquad\qquad\qquad x - 1 = \pm\sqrt{-1}$
$\qquad\qquad\qquad\qquad\qquad\qquad\quad x = 1 \pm i$

So the roots are $x = 2, 1 + i, 1 - i$. As you can see, the complex roots are a conjugate pair.

N2.2 Geometrical implications of complex numbers

Complex conjugates on the Argand diagram

When plotted as vectors on the Argand diagram, a complex number z and its conjugate $\bar{z}$ are reflections in the real axis.

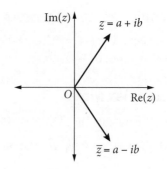

The vector $-z$

The vectors z and $-z$ are the same length but in opposite directions.

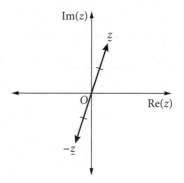

The parallelogram rule for adding and subtracting vectors

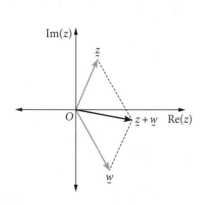

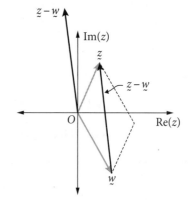

The vector $z + w$ is the diagonal of the parallelogram where z and w are adjacent sides.

The vector $z - w$ is the diagonal of the parallelogram from the head of w to the head of z, the geometrical 'difference' between the 2 vectors.

Multiplying complex numbers by a constant

Multiplying a complex number by a constant (scalar) will change the length of its vector. For example, the vector $2v$ is twice the length of v.

Multiplying numbers on the complex plane

If $z_1 = r_1(\cos \theta_1 + i \sin \theta_1)$ and $z_2 = r_2(\cos \theta_2 + i \sin \theta_2)$ are 2 complex numbers, then their product is

$$z_1 z_2 = r_1 r_2 (\cos (\theta_1 + \theta_2) + i \sin (\theta_1 + \theta_2)).$$

Geometrically, the product of 2 vectors z and w has length $|z||w|$ and argument $\arg z + \arg w$. So if a vector z is multiplied by a vector w, then its modulus is increased by the factor $|w|$ and its argument is rotated **anticlockwise** by $\arg w$.

Example 14

If $z = 3\left(\cos\dfrac{\pi}{3} + i\sin\dfrac{\pi}{3}\right)$ and $w = 2\left(\cos\dfrac{\pi}{6} + i\sin\dfrac{\pi}{6}\right)$, plot z and zw on an Argand diagram.

Solution

Multiply the moduli:

$$|z||w| = 3 \times 2$$
$$= 6$$

Add the arguments:

$$\operatorname{Arg} zw = \frac{\pi}{3} + \frac{\pi}{6}$$
$$= \frac{\pi}{2}$$

So to plot zw, dilate ('stretch') the length of z by a factor of 2 and rotate z by an angle of $\dfrac{\pi}{6}$.

$$zw = 6\left(\cos\frac{\pi}{2} + i\sin\frac{\pi}{2}\right)$$
$$= 6i$$

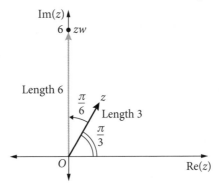

Dividing numbers on the complex plane

If $z_1 = r_1(\cos\theta_1 + i\sin\theta_1)$ and $z_2 = r_2(\cos\theta_2 + i\sin\theta_2)$ are 2 complex numbers, then their **quotient** is

$$\frac{z_1}{z_2} = \frac{r_1}{r_2}\left[\cos(\theta_1 - \theta_2) + i\sin(\theta_1 - \theta_2)\right].$$

Geometrically, this means that the quotient of 2 vectors $\underset{\sim}{z}$ and $\underset{\sim}{w}$ has length $\dfrac{|z|}{|w|}$ and argument $\arg\underset{\sim}{z} - \arg\underset{\sim}{w}$.

So if a vector $\underset{\sim}{z}$ is divided by a vector $\underset{\sim}{w}$, then its modulus is decreased by the factor $|w|$ and its argument is rotated **clockwise** by $\arg\underset{\sim}{w}$.

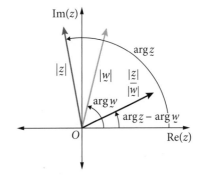

Multiplying and dividing by i

Multiplication by i is equivalent to a rotation of $\dfrac{\pi}{2}$ anticlockwise on the complex plane.

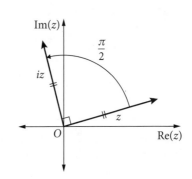

9780170459266

Division by i or multiplication by $-i$ are equivalent to a rotation of $-\dfrac{\pi}{2}$ (that is, $\dfrac{\pi}{2}$ anticlockwise) on the complex plane.

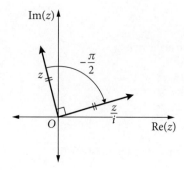 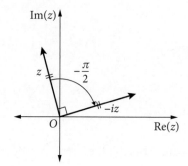

Division by i and multiplication by $-i$ give the same result:

$$\frac{z}{i} = \frac{z}{i} \times \frac{i}{i} = \frac{iz}{-1} = -iz.$$

Example 15

Given $z = 1(\cos\theta + i\sin\theta)$ in the diagram, plot each complex number below on an Argand diagram.

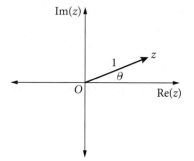

a iz **b** $\dfrac{z}{i}$

c z^2 **d** $\dfrac{1}{z}$

Solution

a $iz = \cos\left(\theta + \dfrac{\pi}{2}\right) + i\sin\left(\theta + \dfrac{\pi}{2}\right)$ **b** $\dfrac{z}{i} = \cos\left(\theta - \dfrac{\pi}{2}\right) + i\sin\left(\theta - \dfrac{\pi}{2}\right)$

c $z^2 = \cos(2\theta) + i\sin(2\theta)$ **d** $\dfrac{1}{z} = \cos(-\theta) + i\sin(-\theta)$

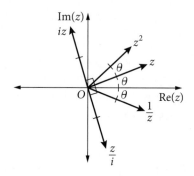

Cube roots of unity

The cube roots of unity (where unity means 1) are the solutions to the equation $z^3 = 1$.

The 3 roots are:

$$z = 1 \text{ or } z = \frac{-1 \pm i\sqrt{3}}{2}, \text{ 1 real and 2 complex conjugates.}$$

In polar form, these roots are:

$$z = 1, \ z = \cos\frac{2\pi}{3} + i\sin\frac{2\pi}{3}, \ z = \cos\left(-\frac{2\pi}{3}\right) + i\sin\left(-\frac{2\pi}{3}\right).$$

When graphed on the complex plane, the roots are equally spaced around the origin.

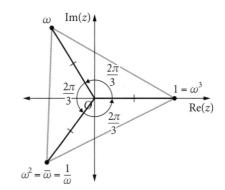

$z = 1$ is on the $\text{Re}(z)$ axis.

Let $\omega = \cos\frac{2\pi}{3} + i\sin\frac{2\pi}{3}$ be the first complex cube root.

Then the following properties hold:

1. $\omega^2 = \bar{\omega} = \dfrac{1}{\omega} = \cos\left(-\frac{2\pi}{3}\right) + i\sin\left(-\frac{2\pi}{3}\right)$

   (ω^2 is the other complex root)

2. $\omega^2 + \omega + 1 = 0$
3. $\omega^3 = 1$

Sum and difference of 2 cubes

These 2 formulas are not on the HSC exam reference sheet but are handy to know for Maths Extension 2 problems involving proofs and complex numbers.

$$a^3 + b^3 = (a + b)(a^2 - ab + b^2)$$

$$a^3 - b^3 = (a - b)(a^2 + ab + b^2)$$

Roots of unity

More generally, the solutions to an equation of the form $z^n = 1$ (where $n \in \mathbb{N}$) are called **roots of unity**.

$z^n = 1$ has n solutions on the complex plane. $z = 1$ is one of the solutions, a real solution.

1. When plotted on the complex plane, the nth roots of unity $z_1, z_2, z_3, z_4, \ldots, z_n$ are equally spaced $\left(\dfrac{2\pi}{n}\right)$ around the origin.

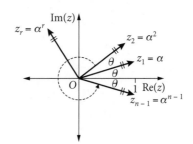

2. The complex roots come in conjugate pairs.

3. The roots $z_1, z_2, z_3, z_4, \ldots, z_n$ form a regular polygon and the vector sum is zero:

   $$z_1 + z_2 + z_3 + z_4 + \cdots + z_n = 0.$$

   Therefore the sum of the roots is 0. (Exception: when $n = 1$, the only root is 1 so the sum is also 1.)

4. Note also that if $z_1 = \alpha, z_2 = \alpha^2, z_3 = \alpha^3 \cdots z_n = \alpha^n = 1$, then $z_1 + z_2 + z_3 + z_4 + \cdots + z_n = 0$ can be written as $1 + \alpha + \alpha^2 + \alpha^3 + \cdots + \alpha^{n-1} = 0$.

5. Note the conjugate pairs: $\bar{\alpha} = \alpha^{n-1}, \ \overline{\alpha^2} = \alpha^{n-2}, \ \overline{\alpha^3} = \alpha^{n-3}, \ \ldots$

Example 16

Find the complex roots of $z^7 = 1$ and show them on the complex plane.

Solution

Solving $z^7 = 1$ algebraically in polar form:

Let $z = r(\cos\theta + i\sin\theta)$ for $-\pi < \theta \le \pi$.

$$\text{LHS} = z^7 = r^7(\cos 7\theta + i\sin 7\theta) \quad \text{(using De Moivre's theorem)}$$

$$\text{RHS} = 1 = 1(\cos 0 + i\sin 0)$$

Then equating we have: $r^7(\cos 7\theta + i\sin 7\theta) = 1(\cos 0 + i\sin 0)$.

$\therefore\ r = 1$ and $\cos 7\theta = \cos 0$ for $-7\pi < 7\theta \le 7\pi$

$\cos 7\theta = 1$

Solving: $7\theta = 0, \pm 2\pi, \pm 4\pi, \pm 6\pi$

$$\therefore\ \theta = 0, \pm\frac{2\pi}{7}, \pm\frac{4\pi}{7}, \pm\frac{6\pi}{7}.$$

For convenience, use the abbreviation cis θ for $\cos\theta + i\sin\theta$.

Therefore, the 7 solutions are:

$$z = \operatorname{cis} 0,\ \operatorname{cis}\frac{2\pi}{7},\ \operatorname{cis}\frac{4\pi}{7},\ \operatorname{cis}\frac{6\pi}{7},\ \operatorname{cis}\left(-\frac{6\pi}{7}\right),\ \operatorname{cis}\left(-\frac{4\pi}{7}\right),\ \operatorname{cis}\left(-\frac{2\pi}{7}\right)$$

$$z = 1,\ \operatorname{cis}\frac{2\pi}{7},\ \operatorname{cis}\frac{4\pi}{7},\ \operatorname{cis}\frac{6\pi}{7},\ \operatorname{cis}\left(-\frac{6\pi}{7}\right),\ \operatorname{cis}\left(-\frac{4\pi}{7}\right),\ \operatorname{cis}\left(-\frac{2\pi}{7}\right)$$

Plotting these, we see that the roots are equally spaced around the Argand diagram, starting at $z = 1$.

Geometrically, we could bypass the algebra and use the pattern developed above to plot 7 equally spaced roots, $\dfrac{2\pi}{7}$ apart around $z = 1$. Joining the roots forms a regular septagon (polygon with 7 sides).

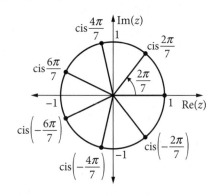

Example 17

ω is a complex cube root of unity.

a   Show that $\omega^2 + \omega + 1 = 0$.

b   Simplify $(1 + 2\omega + 3\omega^2)(1 + 2\omega^2 + 3\omega)$.

Solution

a   Using $z^3 = 1$, then

$$z^3 - 1 = 0$$
$$(z - 1)(z^2 + z + 1) = 0$$
$$z = 1 \quad \text{or} \quad z^2 + z + 1 = 0$$

ω is complex so ω satisfies $z^2 + z + 1 = 0$.

$$\therefore\ \omega^2 + \omega + 1 = 0$$

b   Now we can expand and rearrange

$\omega^2 + \omega + 1 = 0$ as $\omega^2 = -\omega - 1$.

$$(1 + 2\omega + 3\omega^2)(1 + 2\omega^2 + 3\omega)$$
$$= 1 + 2\omega^2 + 3\omega + 2\omega + 4\omega^3 + 6\omega^2 + 3\omega^2 + 6\omega^4 + 9\omega^3$$
$$= 1 + 5\omega + 11\omega^2 + 13\omega^3 + 6\omega^4$$
$$= 1 + 5\omega + 11\omega^2 + 13(1) + 6(1)\omega \quad \because\ \omega^3 = 1$$
$$= 14 + 11\omega + 11\omega^2$$
$$= 14 + 11\omega + 11(-\omega - 1)$$
$$= 14 + 11\omega - 11\omega - 11$$
$$= 3$$

Roots of a complex number

The solutions to $z^n = a + ib$ are also equally spaced $\left(\dfrac{2\pi}{n}\right)$ around the origin starting at $z_1 = r(\cos\theta + i\sin\theta)$, where $\theta = \dfrac{\arg(a + ib)}{n}$ and $r = \sqrt[n]{|a + ib|}$.

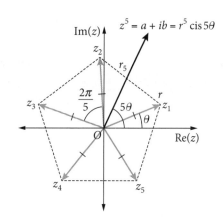

The example of $z^5 = a + ib$ is sketched. The roots z_1, z_2, z_3, z_4, z_5 form the vertices of a regular pentagon.

Using vector addition of the regular polygon, notice that the sum of the roots is zero:

$z_1 + z_2 + z_3 + z_4 + \cdots + z_n = 0$.

However, since $z^n = a + ib$ is complex, the roots *do not* come in conjugate pairs.

Example 18

Solve the equation $z^5 = -\sqrt{3} + i$.

Solution

The 5 roots z_1, z_2, z_3, z_4, z_5 will be equally spaced around the origin starting at z_1.

First find the polar form of $z^5 = -\sqrt{3} + i$.

$$
\begin{aligned}
\left|z^5\right| &= \sqrt{(-\sqrt{3})^2 + 1^2} \\
&= \sqrt{4} \\
&= 2
\end{aligned}
\qquad\qquad
\begin{aligned}
\tan\theta &= \frac{1}{-\sqrt{3}} \\
\theta &= \frac{5\pi}{6}
\end{aligned}
$$

Let $z_1 = r(\cos\varphi + i\sin\varphi)$ be the 1st root.

Then $z_1^5 = r^5(\cos 5\varphi + i\sin 5\varphi) = 2\left(\cos\dfrac{5\pi}{6} + i\sin\dfrac{5\pi}{6}\right)$.

So $r = \sqrt[5]{2}$ and $5\varphi = \dfrac{5\pi}{6}$ so $\varphi = \dfrac{\pi}{6}$.

Therefore $z_1 = \sqrt[5]{2}\left(\cos\dfrac{\pi}{6} + i\sin\dfrac{\pi}{6}\right)$. The other 4 roots will be equally spaced from z_1.

Dividing 2π by 5 we have a spacing of $\dfrac{2\pi}{5}$. So, the other 4 roots are:

$$
\begin{aligned}
z_2 &= \sqrt[5]{2}\left[\cos\left(\frac{\pi}{6} + \frac{2\pi}{5}\right) + i\sin\left(\frac{\pi}{6} + \frac{2\pi}{5}\right)\right] \\
&= \sqrt[5]{2}\cos\frac{17\pi}{30} + i\sin\frac{17\pi}{30}
\end{aligned}
$$

$$
\begin{aligned}
z_4 &= \sqrt[5]{2}\left[\cos\left(\frac{\pi}{6} + \frac{6\pi}{5}\right) + i\sin\left(\frac{\pi}{6} + \frac{6\pi}{5}\right)\right] \\
&= \sqrt[5]{2}\left(\cos\frac{41\pi}{30} + i\sin\frac{41\pi}{30}\right) \\
&= \sqrt[5]{2}\left[\cos\left(-\frac{19\pi}{30}\right) + i\sin\left(-\frac{19\pi}{30}\right)\right] \quad \text{(using the principal argument)}
\end{aligned}
$$

$$
\begin{aligned}
z_3 &= \sqrt[5]{2}\left[\cos\left(\frac{\pi}{6} + \frac{4\pi}{5}\right) + i\sin\left(\frac{\pi}{6} + \frac{4\pi}{5}\right)\right] \\
&= \sqrt[5]{2}\left(\cos\frac{29\pi}{30} + i\sin\frac{29\pi}{30}\right)
\end{aligned}
$$

$$
\begin{aligned}
z_5 &= \sqrt[5]{2}\left[\cos\left(\frac{\pi}{6} + \frac{8\pi}{5}\right) + i\sin\left(\frac{\pi}{6} + \frac{8\pi}{5}\right)\right] \\
&= \sqrt[5]{2}\left(\cos\frac{53\pi}{30} + i\sin\frac{53\pi}{30}\right) \\
&= \sqrt[5]{2}\left[\cos\left(-\frac{7\pi}{30}\right) + i\sin\left(-\frac{7\pi}{30}\right)\right] \quad \text{(using the principal argument)}
\end{aligned}
$$

Note that the roots are equally spaced around

$$z_1 = \sqrt[5]{2}\left(\cos\frac{\pi}{6} + i\sin\frac{\pi}{6}\right).$$

Also note that $z_1 + z_2 + z_3 + z_4 + z_5 = 0$ and that since $-\sqrt{3} + i$ is not real, the roots do not come in conjugate pairs.

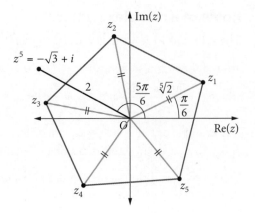

Modulus and argument of $z - z_1$

Given a variable point z and a fixed point z_1 in the complex plane, then:

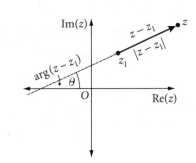

- $|z - z_1|$ is the distance from z to z_1.

- $\arg(z - z_1)$ is the angle between the vector $z - z_1$ and the positive x-axis.

Curves and regions on the number plane

Curves and regions on the complex plane are sets of points representing complex numbers described by a certain rule or condition placed on the variable complex number z. The set of points is often called the **locus** of z.

We can graph a locus in the complex plane **algebraically**, by first deriving the Cartesian equation, or **geometrically** by using the definitions of modulus and argument.

Using the algebraic approach, we let $z = x + iy$, $\text{Re}(z) = x$ and $\text{Im}(z) = y$.

Recall that $|z| = r = \sqrt{x^2 + y^2}$ and $\text{Arg } z$ is the principal argument.

Example 19

Graph each equation on the complex plane.

a $|z + 1 - 2i| = 1$

b $|z - 2| = |z - 2i|$

Solution

a **Algebraic method**

Let $z = x + iy, x, y \in \mathbb{R}$.

We can write $|z + 1 - 2i| = 1$ as

$$|x + iy + 1 - 2i| = 1$$

$$|x + 1 + i(y - 2)| = 1$$

$$\sqrt{(x + 1)^2 + (y - 2)^2} = 1$$

$$(x + 1)^2 + (y - 2)^2 = 1$$

The equation of a circle, with centre $(-1, 2)$, and radius $\sqrt{1} = 1$.

Geometrical method

We can write $|z + 1 - 2i| = 1$ as $|z - (-1 + 2i)| = 1$.

This means the distance of z from $(-1 + 2i)$ is 1 unit.

This is a circle, centre $(-1 + 2i)$ and radius 1.

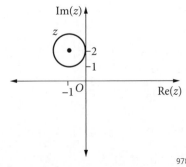

9780170459266

b Algebraic method

Let $z = x + iy, x, y \in \mathbb{R}$.

Then $|z - 2| = |z - 2i|$ becomes

$$|(x - 2) + iy| = |x + (y - 2i)|$$
$$\sqrt{(x - 2)^2 + y^2} = \sqrt{x^2 + (y - 2)^2}$$
$$(x - 2)^2 + y^2 = x^2 + (y - 2)^2$$
$$x^2 - 4x + 4 + y^2 = x^2 + y^2 - 4y + 4$$

So $y = x$; a straight line with gradient 1 going through the origin.

Geometrical method

Geometrically, $|z - 2| = |z - 2i|$ means that z is equidistant from both 2 and $2i$, that is, the perpendicular bisector, $y = x$.

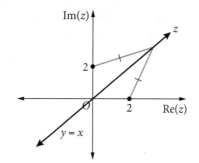

Example 20

Graph each equation on the complex plane.

a $\arg z = \dfrac{\pi}{3}$

b $\arg\left[z - (1 + i)\right] = -\dfrac{\pi}{6}$

c $\arg\left[z - (3 + i)\right] = \arg\left[z - (1 + 3i)\right]$

d $\arg(z - 3) - \arg(z + 3) = \dfrac{\pi}{2}$

Solution

a $\arg z = \dfrac{\pi}{3}$

The complex number z is the vector (or ray) from O at an angle of $\dfrac{\pi}{3}$.

$\arg 0$ is undefined so we draw an open circle there to indicate that it is not part of the graph.

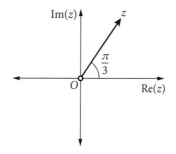

b $\arg\left[z - (1 + i)\right] = -\dfrac{\pi}{6}$

The complex number z is the vector (or ray) from $1 + i$ at an angle of $-\dfrac{\pi}{6}$.

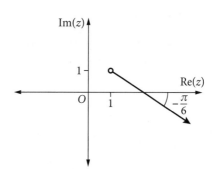

c The arguments are equal so the vectors $z - (3 + i)$ and $z - (1 + 3i)$ must be running in the same direction. They have a common point z, so the locus of $\arg[z - (3 + i)] = [z - (1 + 3i)]$ must be points on the line through $3 + i$ and $1 + 3i$. The solution has 2 sections, excluding the points between $3 + i$ and $1 + 3i$.

Note: The in-between points must be excluded. If z was a point between $3 + i$ and $1 + 3i$, then the vectors $z - (3 + i)$ and $z - (1 + 3i)$ would be running in *opposite* directions.

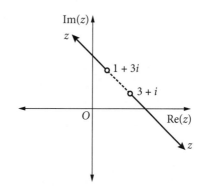

d $\arg(z - 3) - \arg(z + 3) = \dfrac{\pi}{2}$

The arguments differ by 90° but they have a common
point z. It uses the theorem that in a triangle the exterior
angle equals the sum of the 2 interior opposite angles.
The solution is a semicircle with diameter between
3 and −3 because the angle in a semicircle is 90°.

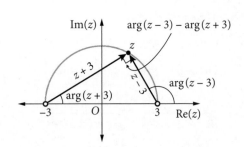

Example 21

Graph each inequality (region) on the complex plane.

a $-\dfrac{\pi}{4} \le \arg z < \dfrac{\pi}{3}$ **b** $\dfrac{1}{2} < |z - 2| \le 1$ **c** $\mathrm{Re}(z) > \mathrm{Im}(z) + 1$

Solution

a $-\dfrac{\pi}{4} \le \arg z < \dfrac{\pi}{3}$ is the region between the 2 vectors

from O with arguments $-\dfrac{\pi}{4}$ and $\dfrac{\pi}{3}$.

Note the dotted vector since $\arg z = \dfrac{\pi}{3}$ is not included.

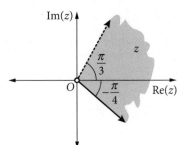

b $\dfrac{1}{2} < |z - 2| \le 1$ represents the region between the

2 concentric circles centred on 2 with radii $\dfrac{1}{2}$ and 1.

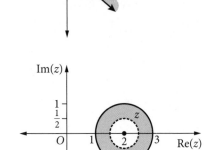

c $\mathrm{Re}(z) > \mathrm{Im}(z) + 1$

Using a Cartesian approach, we can say $x > y + 1$,
or rearranging, $y < x - 1$.

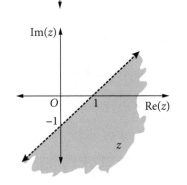

Practice set 1

Multiple-choice questions

Solutions start on page 91.

Question 1 〇〇●

What is the value of i^{59}?

A i **B** $-i$ **C** -1 **D** 1

Question 2 〇●●

Find the value of the expression $\dfrac{3-i}{1+i}$, in simplest form.

A $1 - 2i$ **B** $-1 - 2i$ **C** $1 + 2i$ **D** $-1 + 2i$

Question 3 〇●●

What is the modulus of $\sqrt{7} - 2i$?

A $\sqrt{3}$ **B** 3 **C** $\sqrt{11}$ **D** 11

Question 4 〇〇●

What is the argument of $3\left(\cos\dfrac{\pi}{6} - i\sin\dfrac{\pi}{3}\right)$?

A $-\dfrac{\pi}{3}$ **B** $-\dfrac{\pi}{4}$

C $-\dfrac{\pi}{6}$ **D** 0

Question 5 ●●●

If $z = 5 - 4i(3 - 2i) + 8$, what is $\mathrm{Re}(z)$?

A -20 **B** -12

C 5 **D** 21

Question 6 〇〇●

Find $-\sqrt{2} + i\sqrt{6}$ in polar form.

A $2\left(\cos\dfrac{2\pi}{3} + i\sin\dfrac{2\pi}{3}\right)$ **B** $2\sqrt{2}\left(\cos\dfrac{2\pi}{3} - i\sin\dfrac{2\pi}{3}\right)$

C $2\left[\cos\left(-\dfrac{2\pi}{3}\right) + i\sin\left(-\dfrac{2\pi}{3}\right)\right]$ **D** $2\sqrt{2}\left(\cos\dfrac{2\pi}{3} + i\sin\dfrac{2\pi}{3}\right)$

Question 7 〇〇●

If $(a + ib)^2 = 3 - 4i$, what could be the values of a and b?

A $a = \sqrt{3}, b = 2$ **B** $a = -2, b = -1$

C $a = 1, b = 2$ **D** $a = 2, b = -1$

Question 8 🔘🔘⬛

Simplify $\dfrac{3\left(\cos\dfrac{\pi}{6} + i\sin\dfrac{\pi}{6}\right)}{\sqrt{3}\left[\cos\left(-\dfrac{2\pi}{3}\right) + i\sin\left(-\dfrac{2\pi}{3}\right)\right]}$.

A $\sqrt{3}\left[\cos\left(-\dfrac{\pi}{2}\right) + i\sin\left(-\dfrac{\pi}{2}\right)\right]$

B $\dfrac{1}{\sqrt{3}}\left(\cos\dfrac{\pi}{2} + i\sin\dfrac{\pi}{2}\right)$

C $\sqrt{3}\left(\cos\dfrac{5\pi}{6} + i\sin\dfrac{5\pi}{6}\right)$

D $\dfrac{1}{\sqrt{3}}\left[\cos\left(-\dfrac{5\pi}{6}\right) + i\sin\left(-\dfrac{5\pi}{6}\right)\right]$

Question 9 🔘🔘⬛

Simplify $(e^{\frac{2\pi i}{5}})^8$.

A $e^{\frac{\pi i}{5}}$

B $e^{-\frac{\pi i}{5}}$

C $e^{\frac{4\pi i}{5}}$

D $e^{-\frac{4\pi i}{5}}$

Question 10 🔘🔘⬛

Which equation has $1 - i\sqrt{3}$ as a root?

A $z^2 - 2z + 4 = 0$

B $z^2 + 2z + 4 = 0$

C $z^2 + 4z + 2 = 0$

D $z^2 - 4z + 2 = 0$

Question 11 🔘🔘⬛

What is the equation of the line shown on the Argand diagram?

A $\mathrm{Re}(z) = \mathrm{Im}(z)$

B $\mathrm{Re}(z) + \mathrm{Im}(z) = 1$

C $\mathrm{Re}(z) + \mathrm{Im}(z) = -1$

D $\mathrm{Re}(z) + \mathrm{Im}(z) = 0$

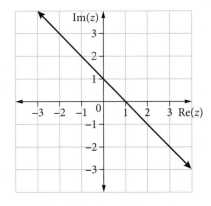

Question 12 🔘🔘🔘

Simplify $\dfrac{1}{(1 - i)^3}$.

A $-2 - 2i$

B $-2 + 2i$

C $-\dfrac{1}{4} - \dfrac{i}{4}$

D $-\dfrac{1}{4} + \dfrac{i}{4}$

Question 13 🔘🔘⬛

What is the complex number z graphed on the Argand diagram?

A $2\left(\cos\dfrac{\pi}{6} + i\sin\dfrac{\pi}{6}\right)$

B $4\left(\cos\dfrac{\pi}{6} - i\sin\dfrac{\pi}{6}\right)$

C $2\left(\cos\dfrac{\pi}{3} + i\sin\dfrac{\pi}{3}\right)$

D $4\left(\cos\dfrac{\pi}{3} - i\sin\dfrac{\pi}{3}\right)$

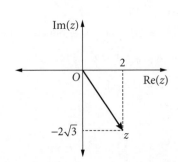

Question 14 ●●●

Which equation below has $\cos\left(-\dfrac{\pi}{6}\right) + i\sin\left(-\dfrac{\pi}{6}\right)$ as a root?

A $z^3 = 1$ **B** $z^6 = 1$ **C** $z^3 = -1$ **D** $z^6 = -1$

Question 15 ●●●

Which equation describes the circle shown on the complex plane?

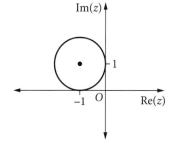

A $|z - 1 - i| = 1$

B $|z + 1 - i| = 1$

C $|z - 1 + i| = 1$

D $|z + 1 + i| = 1$

Question 16 ●●●

Which expression is a factor of the polynomial $z^2 - 6z + 13$?

A $z - 3 - 2i$ **B** $z - 2 - 3i$ **C** $z + 3 - 2i$ **D** $z + 2 - 3i$

Question 17 ●●●

The complex numbers u, v, w, z are graphed on an Argand diagram.

Which of the following equations is true?

A $v = u^2$

B $z = \bar{w}$

C $v = iw$

D $z = -v$

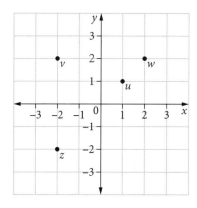

Question 18 ●●●

Simplify $\left|\dfrac{(2 - i)(\sqrt{2} + i\sqrt{2})}{(3 + 4i)}\right|$.

A $\dfrac{2}{\sqrt{5}}$ **B** $\dfrac{2}{5}$ **C** $\dfrac{4}{\sqrt{5}}$ **D** $\dfrac{4}{5}$

Question 19 ●●●

If $(2 - 3i)(x + iy) = 12 - 5i$, where x and y are real, then what are the values of x and y?

A $x = -3, y = -2$ **B** $x = -3, y = 2$ **C** $x = 3, y = -2$ **D** $x = 3, y = 2$

Question 20 ●●●

If $UVWZ$ is a parallelogram in the complex plane and U, V, W, Z represent the complex numbers u, v, w, z, which of the following is true?

A $v - w = w - z$ **B** $v - z = w - u$ **C** $z - u = w - v$ **D** $v - u = z - w$

PRACTICE SET 1

Practice set 2

Short-answer questions

Solutions start on page 92.

Question 1 (6 marks) ●●○

a Find the modulus and argument of $-3 + i\sqrt{3}$.   2 marks

b Express $5\left(i\cos\dfrac{\pi}{3} - \sin\dfrac{\pi}{6}\right)$:

    i in Cartesian form   2 marks

    ii in polar form   2 marks

Question 2 (3 marks) ●●○

Find $\text{Re}(z)$ and $\text{Im}(z)$ if $z = \dfrac{a + ib}{a - ib}$, where a and b are real.   3 marks

Question 3 (4 marks) ●●○

Use De Moivre's theorem to express each expression in polar form.

a $\left(\cos\dfrac{5\pi}{6} - i\sin\dfrac{5\pi}{6}\right)^{12}$ 2 marks

b $\dfrac{1}{\left[\sqrt{2}\left(\cos\dfrac{\pi}{4} + i\sin\dfrac{\pi}{4}\right)\right]^{7}}$ 2 marks

Question 4 (3 marks) ●●○

Simplify $\dfrac{2e^{i\frac{\pi}{12}}}{ie^{i\frac{\pi}{3}}}$ in Cartesian form.   3 marks

Question 5 (5 marks) ●●●

Given that $z_1 = x + iy$ and $z_2 = u + iv$, where $x, y, u, v \in \mathbb{R}$, prove each identity.

a $|z_1|^2 = z_1\overline{z_1}$ 2 marks

b $\left|\dfrac{z_1}{z_2}\right| = \dfrac{|z_1|}{|z_2|}$ 3 marks

Question 6 (5 marks) ●●○

Simplify each expression in polar form.

a $\dfrac{1}{\left(\sqrt{3} - i\right)^2}$ 2 marks

b $(-1 + i)^2\left(1 - i\sqrt{3}\right)^7$ 3 marks

Question 7 (4 marks) ●●○

Solve each equation.

a $z^2 + 4z + 7 = 0$ 2 marks

b $z^2 + 3iz - 2 = 0$ 2 marks

Question 8 (3 marks) ●●●

If $2 - 3i$ is a root of $z^2 + Az + B = 0$, where A and B are real, find the values of A and B.   3 marks

Question 9 (6 marks) ●●●

Solve each equation.

a $z^7 - 1 = 0$ 3 marks

b $z^3 = -8$ 3 marks

Question 10 (6 marks) ●●●

Factorise $z^3 - z - 6$ over:

a the real plane.   3 marks

b the complex plane.   3 marks

Question 11 (5 marks) ●●●

a Prove that   2 marks

$$(\cos\theta_1 + i\sin\theta_1)(\cos\theta_2 + i\sin\theta_2) = \cos(\theta_1 + \theta_2) + i\sin(\theta_1 + \theta_2).$$

b Using part **a**, or otherwise, prove by mathematical induction that for all $n \in \mathbb{N}$:   3 marks

$$(\cos\theta_1 + i\sin\theta_1)(\cos\theta_2 + i\sin\theta_2) \cdots (\cos\theta_n + i\sin\theta_n)$$
$$= \cos(\theta_1 + \theta_2 + \cdots + \theta_n) + i\sin(\theta_1 + \theta_2 + \cdots + \theta_n)$$

Question 12 (8 marks) ●●●

Graph each equation on the complex plane.

a $|z - i| = 2$ 2 marks

b $|z - 2| = |z + 4i|$ 2 marks

c $\arg(z + 3) = \dfrac{\pi}{4}$ 2 marks

d $\arg(z - 1 - i) - \arg(z - 2 - 5i) = \pi$ 2 marks

Question 13 (3 marks) ●●●

Find the monic polynomial equation of minimum degree with real coefficients with roots $1 - 2i$ and $2 + i\sqrt{2}$.   3 marks

Question 14 (10 marks) ●●●

a Let $z = \cos\theta + i\sin\theta$.

Simplify each expression.

i $z^n + \dfrac{1}{z^n}$ 2 marks

ii $z^n - \dfrac{1}{z^n}$ 2 marks

b Expand $\left(e^{i\theta} + \dfrac{1}{e^{i\theta}}\right)^5$ 2 marks

c Hence, or otherwise, evaluate $\displaystyle\int_0^{\frac{\pi}{3}} \cos^5\theta \, d\theta$.   4 marks

Question 15 (7 marks) ●●●

a   By expanding $(\cos\theta + i\sin\theta)^4$ in 2 different ways, show that   3 marks

$$\cos 4\theta = 8\cos^4\theta - 8\cos^2\theta + 1.$$

b   If $x = \cos\theta$, find 4 unique solutions to the equation   4 marks

$$16x^4 - 16x^2 + 1 = 0.$$

Question 16 (6 marks) ●●

a   Find $\sqrt{-24 + 10i}$.   3 marks

b   Hence, or otherwise, solve $z^2 - (5 + 3i)z + 10 + 5i = 0$.   3 marks

Question 17 (2 marks) ●●●

Write the equation describing the set of points on the Argand diagram: 2 marks

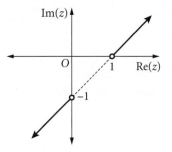

Question 18 (3 marks) ●●●

The complex numbers z_1 and z_2 corresponding with the points Z_1 and Z_2, respectively, are sketched on the Argand plane. It is known that $|z_1| = |z_2|$ and $\angle Z_1OZ_2 = \dfrac{\pi}{6}$.

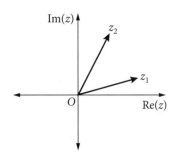

The points $OZ_1Z_3Z_2$, where Z_3 represents z_3, form a rhombus.

Show that $z_3 = \left(1 + e^{\frac{i\pi}{6}}\right)z_1$.   3 marks

Question 19 (3 marks) ●●●

If ω is a complex cube root of unity, simplify $\dfrac{1}{1 + \omega} + \dfrac{1}{\omega + \omega^2} + \dfrac{1}{\omega^2 + 1}$.   3 marks

Question 20 (7 marks) ●●●

ρ is a complex nth root of unity.

a   Explain why $1 + \rho + \rho^2 + \rho^3 + \cdots + \rho^{n-1} = 0$.   3 marks

b   Show that $\left|\rho^n - \rho^{n-1}\right| = \sqrt{2}\sqrt{1 - \cos\left(\dfrac{2\pi}{n}\right)}$.   4 marks

Practice set 1

Worked solutions

1 B

$$i^{59} = \frac{i^{60}}{i} = \frac{(i^4)^{15}}{i} = \frac{1}{i} = i^3 = -i$$

2 A

$$\frac{3-i}{1+i} \times \frac{1-i}{1-i} = \frac{3-3i-i-1}{1+1} = \frac{2-4i}{2} = 1-2i$$

3 C

$$\left|\sqrt{7} - 2i\right| = \sqrt{(\sqrt{7})^2 + (-2)^2} = \sqrt{7+4} = \sqrt{11}$$

4 B

$$3\left(\cos\frac{\pi}{6} - i\sin\frac{\pi}{3}\right) = 3\left(\frac{\sqrt{3}}{2} - i\frac{\sqrt{3}}{2}\right)$$
$$= \frac{3\sqrt{3}}{2}(1-i)$$
$$= \frac{3\sqrt{3}}{2}\left[\cos\left(-\frac{\pi}{4}\right) + i\sin\left(-\frac{\pi}{4}\right)\right]$$

5 C

$$z = 5 - 4i(3-2i) + 8$$
$$= 5 - 12i + 8i^2 + 8$$
$$= 5 - 12i$$

$$\text{Re}(z) = 5$$

6 D

$$-\sqrt{2} + i\sqrt{6} = \sqrt{2}(-1 + i\sqrt{3})$$
$$= \sqrt{2} \times 2\left(\cos\frac{2\pi}{3} + i\sin\frac{2\pi}{3}\right)$$
$$= 2\sqrt{2}\left(\cos\frac{2\pi}{3} + i\sin\frac{2\pi}{3}\right)$$

7 D

$$(a+ib)^2 = 3 - 4i$$
$$a^2 - b^2 + 2abi = 3 - 4i$$
$$a^2 - b^2 = 3, \quad ab = -2$$
$$a = 2, b = -1 \quad \text{or} \quad a = -2, b = 1$$

8 C

$$\frac{3\left(\cos\frac{\pi}{6} + i\sin\frac{\pi}{6}\right)}{\sqrt{3}\left[\cos\left(-\frac{2\pi}{3}\right) + i\sin\left(-\frac{2\pi}{3}\right)\right]}$$
$$= \sqrt{3}\left[\cos\left(\frac{\pi}{6} + \frac{2\pi}{3}\right) + i\sin\left(\frac{\pi}{6} + \frac{2\pi}{3}\right)\right]$$
$$= \sqrt{3}\left(\cos\frac{5\pi}{6} + i\sin\frac{5\pi}{6}\right)$$

9 D

$$(e^{\frac{2\pi i}{5}})^8 = e^{\frac{16\pi i}{5}} = e^{-\frac{4\pi i}{5}}$$

10 A

Roots come in conjugate pairs since coefficients are real.

$$(z - (1 - i\sqrt{3}))(z - (1 + i\sqrt{3})) = 0$$
$$z^2 - (1 - i\sqrt{3} + 1 + i\sqrt{3})z + (1 - i\sqrt{3})(1 + i\sqrt{3}) = 0$$
$$z^2 - 2z + 4 = 0$$

11 B

$$y = -x + 1$$

Graph is $x + y = 1$

$$\text{Re}(z) + \text{Im}(z) = 1$$

12 D

$$\frac{1}{(1-i)^3} = \left\{\sqrt{2}\left[\cos\left(-\frac{\pi}{4}\right) + i\sin\left(-\frac{\pi}{4}\right)\right]\right\}^{-3}$$
$$= \frac{1}{2\sqrt{2}}\left(\cos\frac{3\pi}{4} + i\sin\frac{3\pi}{4}\right)$$
$$= \frac{1}{2\sqrt{2}}\left(-\frac{1}{\sqrt{2}} + \frac{i}{\sqrt{2}}\right)$$
$$= -\frac{1}{4} + \frac{i}{4}$$

13 D

$$r = \sqrt{2^2 + (-2\sqrt{3})^2} = \sqrt{16} = 4$$
$$\arg(z) = -\frac{\pi}{3}$$

So $z = 4\left(\cos\frac{\pi}{3} - i\sin\frac{\pi}{3}\right)$.

14 D

$$\left[\cos\left(-\frac{\pi}{6}\right) + i\sin\left(-\frac{\pi}{6}\right)\right]^6 = \cos(-\pi) + i\sin(-\pi)$$
$$= -1 + 0i$$

15 B

Centre $= (-1, 1)$, radius $= 1$
$$|z - (-1 + i)| = 1$$
$$|z + 1 - i| = 1$$

16 A

$$z^2 - 6z + 13 = z^2 - 6z + 9 + 4$$
$$= (z-3)^2 - (2i)^2$$
$$= (z - 3 - 2i)(z - 3 + 2i)$$

17 C

Since v is the same modulus as w and w is

rotated $\dfrac{\pi}{2}$ anticlockwise, then $v = iw$.

18 A

$$\left|\dfrac{(2-i)(\sqrt{2}+i\sqrt{2})}{(3+4i)}\right| = \dfrac{|2-i||\sqrt{2}+i\sqrt{2}|}{|3+4i|}$$

$$= \dfrac{\sqrt{4+1}\sqrt{2+2}}{\sqrt{9+16}}$$

$$= \dfrac{2\sqrt{5}}{5}$$

$$= \dfrac{2}{\sqrt{5}}$$

19 D

$$(2-3i)(x+iy) = 12-5i$$

$$x+iy = \dfrac{12-5i}{2-3i} \times \dfrac{2+3i}{2+3i}$$

$$= \dfrac{24+36i-10i+15}{4+9}$$

$$= \dfrac{39+26i}{13}$$

$$= 3+2i$$

20 C

If $UVWZ$ is a parallelogram, the opposite sides have the same modulus and vectors running in the same direction have equal arguments.

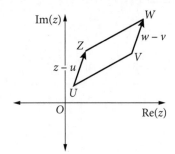

Therefore, $z - u = w - v$ is correct.

Practice set 2

Worked solutions

Question 1

a $\left|-3+i\sqrt{3}\right| = \sqrt{(-3)^2 + (\sqrt{3})^2}$

$$= \sqrt{12}$$

$$= 2\sqrt{3}$$

$$\arg(-3+i\sqrt{3}) = \arg\sqrt{3}(-\sqrt{3}+i)$$

$$= \dfrac{5\pi}{6}$$

b i $5\left(i\cos\dfrac{\pi}{3} - \sin\dfrac{\pi}{6}\right) = 5\left(i\left(\dfrac{1}{2}\right) - \dfrac{1}{2}\right)$

$$= 5\left(-\dfrac{1}{2} + \dfrac{i}{2}\right)$$

$$= -\dfrac{5}{2} + \dfrac{5i}{2}$$

ii $5\left(-\dfrac{1}{2} + \dfrac{i}{2}\right) = \dfrac{5}{\sqrt{2}}\left(\cos\dfrac{3\pi}{4} + i\sin\dfrac{3\pi}{4}\right)$

Question 2

$$z = \dfrac{a+bi}{a-bi} \times \dfrac{a+bi}{a+bi}$$

$$= \dfrac{a^2 - b^2 + 2aib}{a^2 + b^2}$$

$$\text{Re}(z) = \dfrac{a^2 - b^2}{a^2 + b^2}, \; \text{Im}(z) = \dfrac{2ab}{a^2 + b^2}$$

Question 3

a $\left(\cos\dfrac{5\pi}{6} - i\sin\dfrac{5\pi}{6}\right)^{12}$

$$= \left[\cos\left(-\dfrac{5\pi}{6}\right) + i\sin\left(-\dfrac{5\pi}{6}\right)\right]^{12}$$

$$= \cos(-10\pi) + i\sin(-10\pi)$$

$$= \cos 0 + i\sin 0$$

b $\dfrac{1}{\left[\sqrt{2}\left(\cos\dfrac{\pi}{4} + i\sin\dfrac{\pi}{4}\right)\right]^7}$

$$= \dfrac{1}{(\sqrt{2})^7}\left[\cos\left(-\dfrac{7\pi}{4}\right) + i\sin\left(-\dfrac{7\pi}{4}\right)\right]$$

$$= \dfrac{1}{8\sqrt{2}}\left(\cos\dfrac{\pi}{4} + i\sin\dfrac{\pi}{4}\right)$$

Question 4

$$\frac{2e^{i\frac{\pi}{12}}}{ie^{i\frac{\pi}{3}}} = \frac{2e^{i\left(\frac{\pi}{12} - \frac{\pi}{3}\right)}}{e^{i\frac{\pi}{2}}}$$

$$= 2e^{i\left(\frac{\pi}{12} - \frac{\pi}{3} - \frac{\pi}{2}\right)}$$

$$= 2e^{i\left(-\frac{3\pi}{4}\right)}$$

$$= 2\left(-\frac{1}{\sqrt{2}} - \frac{i}{\sqrt{2}}\right)$$

$$= -\sqrt{2} - i\sqrt{2}$$

Question 5

a RTP: $|z_1|^2 = z_1\,\overline{z_1}$

LHS $= |x + iy|^2$

$= \left(\sqrt{x^2 + y^2}\right)^2$

$= x^2 + y^2$

RHS $= z_1\,\overline{z_1}$

$= (x + iy)(x - iy)$

$= x^2 + y^2$

$=$ LHS

$\therefore |z_1|^2 = z_1\,\overline{z_1}$

b RTP: $\left|\dfrac{z_1}{z_2}\right| = \dfrac{|z_1|}{|z_2|}$

LHS $= \left|\dfrac{x + iy}{u + iv} \times \dfrac{u - iv}{u - iv}\right|$

$= \left|\dfrac{xu + yv + i(yu - xv)}{u^2 + v^2}\right|$

$= \sqrt{\dfrac{(xu + yv)^2 + (yu - xv)^2}{(u^2 + v^2)^2}}$

$= \sqrt{\dfrac{x^2u^2 + y^2v^2 + 2xuyv + y^2u^2 + x^2v^2 - 2yuxv}{(u^2 + v^2)^2}}$

$= \sqrt{\dfrac{u^2(x^2 + y^2) + v^2(x^2 + y^2)}{(u^2 + v^2)^2}}$

$= \sqrt{\dfrac{(u^2 + v^2)(x^2 + y^2)}{(u^2 + v^2)(u^2 + v^2)}}$

$= \sqrt{\dfrac{(x^2 + y^2)}{(u^2 + v^2)}}$

RHS $= \dfrac{|x + iy|}{|u + iv|}$

$= \dfrac{\sqrt{x^2 + y^2}}{\sqrt{u^2 + v^2}}$

$= \sqrt{\dfrac{x^2 + y^2}{u^2 + v^2}}$

$=$ LHS

$\therefore \left|\dfrac{z_1}{z_2}\right| = \dfrac{|z_1|}{|z_2|}$

Question 6

a $\dfrac{1}{\left(\sqrt{3} - i\right)^2} = \left\{2\left[\cos\left(-\dfrac{\pi}{6}\right) + i\sin\left(-\dfrac{\pi}{6}\right)\right]\right\}^{-2}$

$= \dfrac{1}{4}\left(\cos\dfrac{\pi}{3} + i\sin\dfrac{\pi}{3}\right)$

b $(-1 + i)^2\left(1 - i\sqrt{3}\right)^7$

$= \left[\sqrt{2}\left(\cos\dfrac{3\pi}{4} + i\sin\dfrac{3\pi}{4}\right)\right]^2 \left\{2\left[\cos\left(-\dfrac{\pi}{3}\right) + i\sin\left(-\dfrac{\pi}{3}\right)\right]\right\}^7$

$= 2\left(\cos\dfrac{3\pi}{2} + i\sin\dfrac{3\pi}{2}\right)2^7\left[\cos\left(-\dfrac{7\pi}{3}\right) + i\sin\left(-\dfrac{7\pi}{3}\right)\right]$

$= 2^8\left[\cos\left(-\dfrac{\pi}{2}\right) + i\sin\left(-\dfrac{\pi}{2}\right)\right]\left[\cos\left(-\dfrac{\pi}{3}\right) + i\sin\left(-\dfrac{\pi}{3}\right)\right]$

$= 2^8\left[\cos\left(-\dfrac{5\pi}{6}\right) + i\sin\left(-\dfrac{5\pi}{6}\right)\right]$

Question 7

a
$$z^2 + 4z + 7 = 0$$
$$z^2 + 4z + 4 + 3 = 0$$
$$(z + 2)^2 - 3i^2 = 0$$
$$\left(z + 2 - i\sqrt{3}\right)\left(z + 2 + i\sqrt{3}\right) = 0$$
$$z = -2 \pm i\sqrt{3}$$

b $z^2 + 3iz - 2 = 0$

$$z = \frac{-3i \pm \sqrt{(3i)^2 - 4(1)(-2)}}{2}$$

$$= \frac{-3i \pm \sqrt{-1}}{2}$$

$$= \frac{-3i \pm i}{2}$$

$$= -i, -2i$$

Question 8

If $2 - 3i$ is a root of $z^2 + Az + B = 0$ and the coefficients are real, then $2 + 3i$ is also a root.

$\therefore z^2 + Az + B$

$= (z - (2 - 3i))(z - (2 + 3i))$

$= z^2 - (2 - 3i + 2 + 3i)z + (2 - 3i)(2 + 3i)$

$= z^2 - 4z + 13$

So $A = -4$, $B = 13$.

Question 9

a $z^7 - 1 = 0$. Roots are equally spaced around O including $1 + 0i$.

$$z_1 = \cos\frac{2\pi}{7} + i\sin\frac{2\pi}{7}, \; z_2 = \cos\frac{4\pi}{7} + i\sin\frac{4\pi}{7}, \; z_3 = \cos\frac{6\pi}{7} + i\sin\frac{6\pi}{7},$$

$$z_4 = \cos\frac{-6\pi}{7} + i\sin\frac{-6\pi}{7}, \; z_5 = \cos\frac{-4\pi}{7} + i\sin\frac{-4\pi}{7}, \; z_6 = \cos\frac{-2\pi}{7} + i\sin\frac{-2\pi}{7}, \; z_7 = 1$$

b $z^3 = -8$. Roots are equally spaced around O, including $-2 + 0i$ since $z = \sqrt[3]{-8} = -2$.

Roots are $z_1 = \cos\frac{\pi}{3} + i\sin\frac{\pi}{3}, \; z_2 = \cos\frac{-\pi}{3} + i\sin\frac{-\pi}{3}, \; z_3 = -2$.

[Roots equally spaced; includes complex conjugates since $z^3 = -8$ has real coefficients.]

Question 10

a $z^3 - z - 6$ has a real root.

Check:

$P(2) = 2^3 - 2 - 6 = 0$.
Therefore, $z - 2$ is a factor.

So $z^3 - z - 6 = (z - 2)(z^2 + Az + B)$.

Equate coefficients:

$-6 = -2B \Rightarrow B = 3$

$0z^2 = -2z^2 + Az^2$

$\quad 0 = -2 + A$

$\quad A = 2$

So $z^3 - z - 6 = (z - 2)(z^2 + 2z + 3)$ over the real plane.

b Factorise $(z^2 + 2z + 3)$.

$$(z^2 + 2z + 3) = (z^2 + 2z + 1 + 2)$$
$$= (z + 1)^2 - (i\sqrt{2})^2$$
$$= (z + 1 - i\sqrt{2})(z + 1 + i\sqrt{2})$$

So $z^3 - z - 6 = (z - 2)(z + 1 - i\sqrt{2})(z + 1 + i\sqrt{2})$ over the complex plane.

Question 11

a RTP: $(\cos\theta_1 + i\sin\theta_1)(\cos\theta_2 + i\sin\theta_2) = \cos(\theta_1 + \theta_2) + i\sin(\theta_1 + \theta_2)$.

Proof:

$$(\cos\theta_1 + i\sin\theta_1)(\cos\theta_2 + i\sin\theta_2) = \cos\theta_1\cos\theta_2 + \cos\theta_1 i\sin\theta_2 + i\sin\theta_1\cos\theta_2 - \sin\theta_1\sin\theta_2$$
$$= (\cos\theta_1\cos\theta_2 - \sin\theta_1\sin\theta_2) + i(\cos\theta_1\sin\theta_2 + \sin\theta_1\cos\theta_2)$$
$$= \cos(\theta_1 + \theta_2) + i\sin(\theta_1 + \theta_2)$$

using the trigonometric identities.

b Let $P(n)$ be the proposition that for $n \in \mathbb{N}$

$(\cos \theta_1 + i \sin \theta_1)(\cos \theta_2 + i \sin \theta_2) \cdots (\cos \theta_n + i \sin \theta_n) = \cos(\theta_1 + \theta_2 + \cdots + \theta_n) + i \sin(\theta_1 + \theta_2 + \cdots + \theta_n)$.

Proof:

Prove true for $P(1)$:

LHS $= (\cos \theta_1 + i \sin \theta_1)$

RHS $= \cos(\theta_1) + i \sin(\theta_1)$

$\therefore$ LHS = RHS

$P(1)$ is true.

Assume $P(k)$ is true for some $k \in \mathbb{N}$, that is,

$(\cos \theta_1 + i \sin \theta_1)(\cos \theta_2 + i \sin \theta_2) \cdots (\cos \theta_k + i \sin \theta_k)$
$= \cos(\theta_1 + \theta_2 + \cdots + \theta_k) + i \sin(\theta_1 + \theta_2 + \cdots + \theta_k)$

RTP: $P(k+1)$ is true, i.e.

$(\cos \theta_1 + i \sin \theta_1)(\cos \theta_2 + i \sin \theta_2) \cdots (\cos \theta_k + i \sin \theta_k)(\cos \theta_{k+1} + i \sin \theta_{k+1})$
$= \cos(\theta_1 + \theta_2 + \cdots + \theta_k + \theta_{k+1}) + i \sin(\theta_1 + \theta_2 + \cdots + \theta_k + \theta_{k+1})$

Proof:

Consider the LHS of $P(k+1)$:

$(\cos \theta_1 + i \sin \theta_1)(\cos \theta_2 + i \sin \theta_2) \cdots (\cos \theta_k + i \sin \theta_k)(\cos \theta_{k+1} + i \sin \theta_{k+1})$
$= (\cos(\theta_1 + \theta_2 + \cdots + \theta_k) + i \sin(\theta_1 + \theta_2 + \cdots + \theta_k))(\cos \theta_{k+1} + i \sin \theta_{k+1})$ 　using $P(k)$

Let $\theta_1 + \theta_2 + \cdots + \theta_k = X$ then

$= (\cos(X) + i \sin(X))(\cos \theta_{k+1} + i \sin \theta_{k+1})$
$= \cos(X + \theta_{k+1}) + i \sin(X + \theta_{k+1})$ 　(from part **a**)
$= \cos(\theta_1 + \theta_2 + \cdots + \theta_k + \theta_{k+1}) + i \sin(\theta_1 + \theta_2 + \cdots + \theta_k + \theta_{k+1})$
$=$ RHS of $P(k+1)$

Therefore $P(k+1)$ is true.

So $P(n)$ is true by mathematical induction.

Question 12

a $|z - i| = 2$

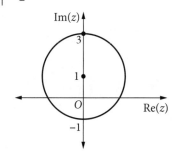

c $\arg(z + 3) = \dfrac{\pi}{4}$

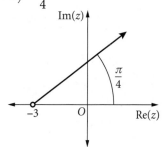

b $|z - 2| = |z + 4i|$

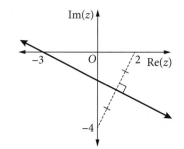

d $\arg(z - 1 - i) - \arg(z - 2 - 5i) = \pi$

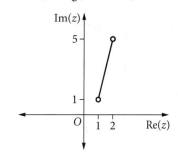

Question 13

Real coefficients so roots must come in conjugate pairs. Roots are $1 \pm 2i$ and $2 \pm i\sqrt{2}$.
Minimum degree 4. Polynomial is:

$$(z-(1+2i))(z-(1-2i))(z-(2+i\sqrt{2}))(z-(2-i\sqrt{2})) = (z^2 - 2z + 5)(z^2 - 4z + 6)$$

So on expansion:

$$z^4 - 6z^3 + 19z^2 - 32z + 30 = 0$$

Question 14

a i $\quad z^n + \dfrac{1}{z^n} = (\cos\theta + i\sin\theta)^n + (\cos\theta + i\sin\theta)^{-n}$

$\qquad\qquad = \cos n\theta + i\sin n\theta + \cos(-n\theta) + i\sin(-n\theta)$

$\qquad\qquad = \cos n\theta + i\sin n\theta + \cos n\theta - i\sin n\theta$

$\qquad\qquad = 2\cos n\theta$

ii $\quad z^n - \dfrac{1}{z^n} = (\cos\theta + i\sin\theta)^n - (\cos\theta + i\sin\theta)^{-n}$

$\qquad\qquad = \cos n\theta + i\sin n\theta - \cos(-n\theta) - i\sin(-n\theta)$

$\qquad\qquad = \cos n\theta + i\sin n\theta - \cos n\theta + i\sin n\theta$

$\qquad\qquad = 2i\sin n\theta$

b $\left(e^{i\theta} + \dfrac{1}{e^{i\theta}}\right)^5 = (e^{i\theta})^5 + 5(e^{i\theta})^4\left(\dfrac{1}{e^{i\theta}}\right)^1 + 10(e^{i\theta})^3\left(\dfrac{1}{e^{i\theta}}\right)^2 + 10(e^{i\theta})^2\left(\dfrac{1}{e^{i\theta}}\right)^3 + 5(e^{i\theta})^1\left(\dfrac{1}{e^{i\theta}}\right)^4 + \left(\dfrac{1}{e^{i\theta}}\right)^5$

$\qquad\qquad = e^{5i\theta} + 5e^{3i\theta} + 10e^{i\theta} + 10e^{-i\theta} + 5e^{-3i\theta} + e^{-5i\theta}$

$\qquad\qquad = \left(e^{5i\theta} + e^{-5i\theta}\right) + 5\left(e^{3i\theta} + e^{-3i\theta}\right) + 10\left(e^{i\theta} + e^{-i\theta}\right)$

c Using part **a**, $z^n + \dfrac{1}{z^n} = 2\cos n\theta$, so

$$\left(e^{i\theta} + \dfrac{1}{e^{i\theta}}\right)^5 = \left(e^{5i\theta} + e^{-5i\theta}\right) + 5\left(e^{3i\theta} + e^{-3i\theta}\right) + 10\left(e^{i\theta} + e^{-i\theta}\right)$$

$$(2\cos\theta)^5 = 2\cos 5\theta + 5(2\cos 3\theta) + 10(2\cos\theta)$$

$$32\cos^5\theta = 2\cos 5\theta + 10\cos 3\theta + 20\cos\theta$$

$$\cos^5\theta = \dfrac{1}{16}\cos 5\theta + \dfrac{5}{16}\cos 3\theta + \dfrac{5}{8}\cos\theta.$$

$$\int_0^{\frac{\pi}{3}}\cos^5\theta\, d\theta = \int_0^{\frac{\pi}{3}} \dfrac{1}{16}\cos 5\theta + \dfrac{5}{16}\cos 3\theta + \dfrac{5}{8}\cos\theta\, d\theta$$

$$= \left[\dfrac{1}{80}\sin 5\theta + \dfrac{5}{48}\sin 3\theta + \dfrac{5}{8}\sin\theta\right]_0^{\frac{\pi}{3}}$$

$$= \dfrac{1}{80}\sin\dfrac{5\pi}{3} + \dfrac{5}{48}\sin\pi + \dfrac{5}{8}\sin\dfrac{\pi}{3} - 0$$

$$= \dfrac{1}{80}\left(-\dfrac{\sqrt{3}}{2}\right) + \dfrac{5}{48}(0) + \dfrac{5}{8}\left(\dfrac{\sqrt{3}}{2}\right)$$

$$= \dfrac{49\sqrt{3}}{160}$$

An alternative approach uses integration by substitution:

$$\int_0^{\frac{\pi}{3}}\cos^5\theta\, d\theta = \int_0^{\frac{\pi}{3}}\cos\theta(1 - \sin^2\theta)^2\, d\theta$$

$$= \int_0^{\frac{\pi}{3}}\cos\theta(1 - 2\sin^2\theta + \sin^4\theta)\, d\theta$$

Substitute $u = \sin\theta$ and solve to obtain the same answer as above.

Question 15

a $\cos 4\theta + i\sin 4\theta = \cos^4\theta + 4\cos^3\theta\, i\sin\theta - 6\cos^2\theta\sin^2\theta - 4\cos\theta\, i\sin^3\theta + \sin^4\theta$

Equating real parts:
$$\begin{aligned}
\cos 4\theta &= \cos^4\theta - 6\cos^2\theta\sin^2\theta + \sin^4\theta \\
&= \cos^4\theta - 6\cos^2\theta(1 - \cos^2\theta) + (1 - \cos^2\theta)^2 \\
&= \cos^4\theta - 6\cos^2\theta + 6\cos^4\theta + 1 - 2\cos^2\theta + \cos^4\theta \\
&= 8\cos^4\theta - 8\cos^2\theta + 1
\end{aligned}$$

b Let $x = \cos\theta$, then $\cos 4\theta = 8x^4 - 8x^2 + 1$.

If $\cos 4\theta = \dfrac{1}{2}$, then

$8x^4 - 8x^2 + 1 = \dfrac{1}{2}$

$\therefore\ 16x^4 - 16x^2 + 1 = 0.$

Solving:

$\cos 4\theta = \dfrac{1}{2}$

$4\theta = \pm\dfrac{\pi}{3},\ \pm\dfrac{5\pi}{3},\ \pm\dfrac{7\pi}{3},\ \pm\dfrac{11\pi}{3},\ \pm\dfrac{13\pi}{3},\ \dots$

$\theta = \pm\dfrac{\pi}{12},\ \pm\dfrac{5\pi}{12},\ \pm\dfrac{7\pi}{12},\ \pm\dfrac{11\pi}{12},\ \pm\dfrac{13\pi}{12},\ \dots$

4 unique solutions are:

$x = \cos\dfrac{\pi}{12},\ \cos\dfrac{5\pi}{12},\ \cos\dfrac{7\pi}{12},\ \cos\dfrac{11\pi}{12}$

Question 16

a Let $\sqrt{-24 + 10i} = a + ib,\ a, b \in \mathbb{R}$

$\therefore\ -24 + 10i = (a + ib)^2$ (equating real and
$-24 + 10i = a^2 - b^2 + 2aib$ imaginary parts)
$-24 = a^2 - b^2$

$10 = 2ab \Rightarrow 5 = ab$

By inspection,

$a = 1, b = 5$ or $a = -1, b = -5$

$\sqrt{-24 + 10i} = \pm(1 + 5i).$

b $z^2 - (5 + 3i)z + 10 + 5i = 0$

$$\begin{aligned}
z &= \frac{5 + 3i \pm \sqrt{(-(5 + 3i))^2 - 4(1)(10 + 5i)}}{2(1)} \\
&= \frac{5 + 3i \pm \sqrt{25 - 9 + 30i - 40 - 20i}}{2} \\
&= \frac{5 + 3i \pm \sqrt{-24 + 10i}}{2} \\
&= \frac{5 + 3i \pm (1 + 5i)}{2} \\
&= 3 + 4i \text{ or } 2 - i
\end{aligned}$$

[Note: coefficients not real so roots aren't conjugate pairs.]

Question 17

The argument from any point on the 2 rays is the same from 1 and from $-i$ so the equation is $\arg(z - 1) = \arg(z + i)$.

Question 18

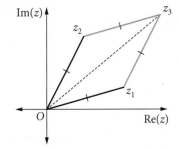

$z_3 = z_1 + z_2$ and

$z_2 = z_1 \times \left(\cos\dfrac{\pi}{6} + i\sin\dfrac{\pi}{6}\right) = z_1 \times e^{\frac{i\pi}{6}}.$

So $z_3 = z_1 + z_1 \times e^{\frac{i\pi}{6}}$

$z_3 = z_1\left(1 + e^{\frac{i\pi}{6}}\right).$

Question 19

If ω is a complex cube root of unity, then $\omega^3 = 1$ and $1 + \omega + \omega^2 = 0$

so $\dfrac{1}{1 + \omega} + \dfrac{1}{\omega + \omega^2} + \dfrac{1}{\omega^2 + 1} = \dfrac{1}{-\omega^2} + \dfrac{1}{-1} + \dfrac{1}{-\omega}$

$ = -\omega - 1 - \omega^2$

$ = -(1 + \omega + \omega^2)$

$ = 0.$

Question 20

a If ρ is a complex nth root of unity, then ρ is a solution to $z^n = 1$, that is $\rho^n = 1$.

Therefore $\rho^n - 1 = 0$ and each root is $\frac{2\pi}{n}$ apart, starting at $(1,0)$,

so the roots are $1, \rho, \rho^2, \rho^3, \ldots, \rho^{n-1}$.

Sum of the roots $= \frac{-b}{a} = 0$ so

$1 + \rho + \rho^2 + \rho^3 + \cdots + \rho^{n-1} = 0$.

[OR
$\rho^n - 1 = (\rho - 1)(\rho^{n-1} + \rho^{n-2} + \rho^{n-3} + \cdots \rho^1 + 1)$
$= 0$

So $1 + \rho + \rho^2 + \rho^3 + \cdots + \rho^{n-1} = 0$
since $\rho \neq 1$ as ρ is complex, not real.]

b

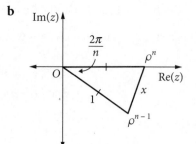

Let x be the distance between roots ρ^{n-1} and ρ^n so using the cosine rule:

$x^2 = \left|\rho^n - \rho^{n-1}\right|^2 = 1^2 + 1^2 - 2(1)(1)\cos\frac{2\pi}{n}$

$x^2 = 2 - 2\cos\left(\frac{2\pi}{n}\right)$

Since $x > 0$

$x = \sqrt{2 - 2\cos\frac{2\pi}{n}}$

$\quad = \sqrt{2\left(1 - \cos\frac{2\pi}{n}\right)}$

$\quad = \sqrt{2}\sqrt{1 - \cos\frac{2\pi}{n}}$.

So $\left|\rho^n - \rho^{n-1}\right| = \sqrt{2}\sqrt{1 - \cos\frac{2\pi}{n}}$.

HSC exam topic grid (2011–2020)

This table shows the coverage of this topic in past HSC exams by question number. The past exams can be downloaded from the NESA website (www.educationstandards.nsw.edu.au) by selecting 'Year 11 – Year 12', 'HSC exam papers'. NESA marking feedback and guidelines can also be found there.

The new Mathematics Extension 2 course was first examined in 2020. For exams before 2020, select 'Year 11 – Year 12', 'Resources archive', 'HSC exam papers archive'.

Euler's formula and exponential form were introduced to the Mathematics Extension 2 course in 2020.

	Cartesian and polar form	Euler's formula and exponential form	De Moivre's theorem and equations	Vectors, roots, curves and regions
2011	2(a), (b), (c)	Introduced in 2020	2(d)	4(a), 6(c)
2012	1, 3, 11(a), 11(d)		11(d), 15(b)	11(b), 12(d)
2013	11(a), 14(b), 15(a)		11(c)	3, 5, 11(e)
2014	4, 11(a)		2, 14(a), 15(b)	8, 11(c)
2015	2, 5, 11(a), 11(b), 12(a)		16(b)(i)	9
2016	10, 11(a), 16(a)		12(c)	4, 5, 16(a), 16(b)
2017	11(a), 16(a)		6, 12(b)	1, 3, 11(c), 13(e)
2018	9, 11(a), 11(d), 13(b)		15(b), 16(c)(iii)	6, 7
2019	1, 8, 11(a), 11(e)		11(e), 16(b)	12(a)
2020 new course	2, 4, 11(a), 14(e)	9, 13(d), 14(a)	4, 11(e)	9

CHAPTER 4
FURTHER INTEGRATION

4

FURTHER INTEGRATION

Integration by substitution

- Let $u = \ldots$
- The substitution may not be given
- Change limits of definite integrals
- Trigonometric substitutions, including t-formulas

Rational functions and partial fractions

- Quadratic denominators that require completing the square
- Linear and quadratic denominators
- Partial fractions: equating coefficients vs substitution
- Often involve logarithmic or inverse trigonometric functions

Integration by parts

- $\int uv' \, dx = uv - \int vu' \, dx$
- Choosing u and v'

Integration by recurrence relations

- Integrals I_n, I_{n-1} and I_0
- Usually involves integration by parts

The standard integrals (also on the HSC exam reference sheet)

$$\int f'(x)\left[f(x)\right]^n dx = \frac{1}{n+1}\left[f(x)\right]^{n+1} + c \text{ where } n \neq -1$$

$$\int f'(x)e^{f(x)} dx = e^{f(x)} + c$$

$$\int \frac{f'(x)}{f(x)} dx = \ln\left|f(x)\right| + c$$

$$\int f'(x)a^{f(x)} dx = \frac{a^{f(x)}}{\ln a} + c$$

$$\int f'(x)\sin f(x) \, dx = -\cos f(x) + c$$

$$\int f'(x)\cos f(x) \, dx = \sin f(x) + c$$

$$\int f'(x)\sec^2 f(x) \, dx = \tan f(x) + c$$

$$\int \frac{f'(x)}{\sqrt{a^2 - \left[f(x)\right]^2}} dx = \sin^{-1}\frac{f(x)}{a} + c$$

$$\int \frac{f'(x)}{a^2 + \left[f(x)\right]^2} dx = \frac{1}{a}\tan^{-1}\frac{f(x)}{a} + c$$

Glossary

integration by parts
A method of integrating a function by splitting it into one function to be differentiated and one function to be integrated:

$$\int u \frac{dv}{dx}\, dx = uv - \int v \frac{du}{dx}\, dx$$

partial fractions
A rational function can be expressed as the sum of smaller fractions called partial fractions that are easier to integrate.

rational function
A function that can be expressed as a fraction $\frac{f(x)}{g(x)}$, such that both the numerator and the denominator are polynomials.

A+ DIGITAL FLASHCARDS
Revise this topic's key terms and concepts by scanning the QR code or typing the URL into your browser.

https://get.ga/a-hsc-maths-ext-2

recurrence relation or recursive formula
A formula or integral that is expressed in terms of itself with a smaller parameter value, for example,

$$\int_0^{\frac{\pi}{2}} \sin^n\theta\, d\theta = \frac{n-1}{n} \int_0^{\frac{\pi}{2}} \sin^{n-2}\theta\, d\theta$$

so this integral involving a power of n can be expressed as an integral involving a power of $n - 2$. That is:

$$I_n = \frac{n-1}{n} I_{n-2}.$$

9780170459266

Topic summary

Further integration (MEX-C1)

Integration by substitution

Integration by substitution is used for integrals of composite functions involving a function and its derivative. We can use the reverse chain rule

$$\int f'(x)\left[f(x)\right]^n dx = \frac{1}{n+1}\left[f(x)\right]^{n+1} + c.$$

Example 1

Find $\int \dfrac{4x}{\sqrt{2x^2 + 1}}\, dx.$

Solution

Let $u = 2x^2 + 1.$

$\dfrac{du}{dx} = 4x$

$du = 4x\, dx$

$dx = \dfrac{du}{4x}$

So $\int \dfrac{4x}{\sqrt{2x^2 + 1}}\, dx = \int \dfrac{4x}{\sqrt{u}} \dfrac{du}{4x}$

$= \int \dfrac{1}{\sqrt{u}} du$

$= \int u^{-\frac{1}{2}} du$

$= \dfrac{1}{\frac{1}{2}} u^{\frac{1}{2}} + c$

$= 2\sqrt{u} + c$

$= 2\sqrt{2x^2 + 1} + c.$

Example 2

Find $\int \dfrac{\sin \sqrt{x}}{\sqrt{x}}\, dx.$

Solution

Let $u = \sqrt{x}.$

$\dfrac{du}{dx} = \dfrac{1}{2\sqrt{x}}$

$du = \dfrac{1}{2\sqrt{x}}\, dx$

$dx = 2\sqrt{x}\, du$

$\int \dfrac{\sin \sqrt{x}}{\sqrt{x}}\, dx = \int \dfrac{\sin u}{\sqrt{x}} 2\sqrt{x}\, du$

$= 2\int \sin u\, du$

$= -2\cos u + c$

$= -2\cos \sqrt{x} + c$

Trigonometric integrals of the form $\int \sin^n x \cos x\, dx$ or $\int \cos^n x \sin x\, dx$ require a simple trigonometric substitution.

Example 3

Find $\int \sin^9 x \cos x\, dx$.

Solution

Let $u = \sin x$.

$\dfrac{du}{dx} = \cos x$

$du = \cos x\, dx$

$dx = \dfrac{du}{\cos x}$

$$\int \sin^9 x \cos x\, dx = \int u^9 \cos x \frac{du}{\cos x}$$
$$= \int u^9\, du$$
$$= \frac{1}{10} u^{10} + c$$
$$= \frac{1}{10} \sin^{10} x + c$$

Example 4

Evaluate $\displaystyle\int_1^4 \frac{1}{(1 + \sqrt{x})^2 \sqrt{x}}\, dx$.

Solution

Let $u = 1 + \sqrt{x}$.

$\dfrac{du}{dx} = \dfrac{1}{2\sqrt{x}}$

$du = \dfrac{1}{2\sqrt{x}}\, dx$

$dx = 2\sqrt{x}\, du$

Change limits of integration:

When $x = 1$, $u = 1 + \sqrt{1} = 2$.

When $x = 4$, $u = 1 + \sqrt{4} = 3$.

$$\int_1^4 \frac{1}{(1 + \sqrt{x})^2 \sqrt{x}}\, dx = \int_2^3 \frac{1}{u^2 \sqrt{x}} 2\sqrt{x}\, du$$
$$= 2\int_2^3 \frac{1}{u^2}\, du$$
$$= 2\left[-\frac{1}{u} \right]_2^3$$
$$= 2\left(-\frac{1}{3} - \left[-\frac{1}{2} \right] \right)$$
$$= 2\left(\frac{1}{6} \right)$$
$$= \frac{1}{3}$$

t-formulas

For integrals of the form $\displaystyle\int \frac{dx}{a\cos x + b\sin x}$, $\displaystyle\int \frac{dx}{a\cos x + b}$ or $\displaystyle\int \frac{dx}{a + b\sin x}$, we can use the *t*-formulas, where $t = \tan\left(\dfrac{x}{2} \right)$:

$$\sin x = \frac{2t}{1 + t^2} \qquad \cos x = \frac{1 - t^2}{1 + t^2} \qquad \tan x = \frac{2t}{1 - t^2} \qquad dx = \frac{2dt}{1 + t^2}.$$

TOPIC SUMMARY

Example 5

Use the substitution $t = \tan\dfrac{\theta}{2}$ to show that $\displaystyle\int_{\frac{\pi}{2}}^{\frac{2\pi}{3}} \dfrac{d\theta}{\sin\theta} = \dfrac{1}{2}\ln 3$.

Solution

$$t = \tan\frac{\theta}{2}$$

$$\frac{dt}{d\theta} = \frac{1}{2}\sec^2\left(\frac{\theta}{2}\right)$$

$$\frac{d\theta}{dt} = \frac{2}{\sec^2\left(\dfrac{\theta}{2}\right)}$$

$$= \frac{2}{1 + \tan^2\left(\dfrac{\theta}{2}\right)}$$

$$= \frac{2}{1 + t^2}$$

$$d\theta = \frac{2dt}{1 + t^2}$$

$$\sin\theta = \frac{2t}{1 + t^2}$$

When $\theta = \dfrac{\pi}{2}$, $t = \tan\dfrac{\pi}{4} = 1$.

When $\theta = \dfrac{2\pi}{3}$, $t = \tan\dfrac{2\pi}{6}$

$$= \tan\frac{\pi}{3}$$

$$= \sqrt{3}.$$

$$\int_{\frac{\pi}{2}}^{\frac{2\pi}{3}} \frac{d\theta}{\sin\theta} = \int_1^{\sqrt{3}} \frac{1 + t^2}{2t}\,\frac{2dt}{1 + t^2}$$

$$= \int_1^{\sqrt{3}} \frac{1}{t}\,dt$$

$$= \Big[\ln|t|\Big]_1^{\sqrt{3}}$$

$$= \ln\sqrt{3} - \ln 1$$

$$= \ln\sqrt{3}$$

$$= \ln 3^{\frac{1}{2}}$$

$$= \frac{1}{2}\ln 3, \text{ as required.}$$

> **Hint**
>
> When the integrand has a sum or difference of 2 squares:
>
> $\displaystyle\int \dfrac{dx}{\sqrt{a^2 + x^2}}$ use the substitution $x = a\tan\theta$
>
> $\displaystyle\int \sqrt{a^2 - x^2}\,dx$ use the substitution $x = a\sin\theta$ or $x = a\cos\theta$
>
> $\displaystyle\int \dfrac{dx}{\sqrt{x^2 - a^2}}$ use the substitution $x = a\sec\theta$

$\sin^2 nx$ and $\cos^2 nx$

$$\sin^2 nx = \frac{1}{2}(1 - \cos 2nx)$$

$$\cos^2 nx = \frac{1}{2}(1 + \cos 2nx)$$

The above trigonometric identities from the Mathematics Extension 1 course are on the HSC exam reference sheet and allow other similar integrations to be quickly determined.

$$\sin^2 x = \frac{1}{2}(1 - \cos 2x) \qquad \text{OR} \qquad \sin^2 2x = \frac{1}{2}(1 - \cos 4x)$$

$$\cos^2 x = \frac{1}{2}(1 + \cos 2x) \qquad \text{OR} \qquad \cos^2 3x = \frac{1}{2}(1 + \cos 6x)$$

Example 6

Evaluate $\displaystyle\int_0^{\frac{\pi}{4}} \cos^2 2x\,dx$.

Solution

$$\int_0^{\frac{\pi}{4}} \cos^2 2x\,dx = \int_0^{\frac{\pi}{4}} \frac{1}{2}\big[1 + \cos 4x\big]\,dx$$

$$= \frac{1}{2}\left[x + \frac{1}{4}\sin 4x\right]_0^{\frac{\pi}{4}}$$

$$= \frac{1}{2}\left[\left(\frac{\pi}{4} + 0\right) - (0 + 0)\right]$$

$$= \frac{\pi}{8}$$

> **Hint**
>
> For integrands that have sin and cos terms, or are of the form $\int \sin^m x\cos^n x\,dx$:
>
> • If both sin and cos have even powers, then use the trigonometric identities (double angle formulae) to expand them into sin and cos terms with single powers, as seen in this example.
>
> • If either sin or cos has an odd power, then substitute $(1 - \sin^2)$ or $(1 - \cos^2)$ and use integration by substitution, as seen in Example 3.

Rational functions with quadratic denominators

A function $\dfrac{f(x)}{g(x)}$, where $f(x)$ and $g(x)$ are polynomials, is called a rational function.

For integrals involving a quadratic denominator, strategies include:

- factorising

- completing the square

- splitting the numerator

- recognising the reverse chain rule $\displaystyle\int f'(x)\left[f(x)\right]^n dx = \dfrac{1}{n+1}\left[f(x)\right]^{n+1} + c$

- recognising an integral that becomes a logarithmic or inverse trigonometric function

$$\int \frac{f'(x)}{f(x)} dx = \ln|f(x)| + c$$

$$\int \frac{f'(x)}{\sqrt{a^2 - \left[f(x)\right]^2}} dx = \sin^{-1}\frac{f(x)}{a} + c \qquad \int \frac{f'(x)}{a^2 + \left[f(x)\right]^2} dx = \frac{1}{a}\tan^{-1}\frac{f(x)}{a} + c$$

TOPIC SUMMARY

Example 7

a Find $\displaystyle\int \frac{1}{\sqrt{3 - 2x - x^2}} dx$.

b Find $\displaystyle\int \frac{2x + 1}{x^2 + 2x + 3} dx$.

Solution

a Complete the square.

$$\int \frac{1}{\sqrt{3 - 2x - x^2}} dx = \int \frac{1}{\sqrt{-(x^2 + 2x + 1) + 4}} dx$$

$$= \int \frac{1}{\sqrt{4 - (x + 1)^2}} dx$$

$$= \sin^{-1}\left(\frac{x + 1}{2}\right) + c$$

recognising $\displaystyle\int \frac{f'(x)}{\sqrt{a^2 - \left[f(x)\right]^2}} dx = \sin^{-1}\frac{f(x)}{a} + c$,

where $f(x) = x + 1$.

b Recognise that the numerator is almost the derivative of the denominator.

$$\int \frac{2x + 1}{x^2 + 2x + 3} dx$$

$$= \int \frac{2x + 2}{x^2 + 2x + 3} - \frac{1}{x^2 + 2x + 3} dx$$

$$= \ln|x^2 + 2x + 3| - \int \frac{1}{(x + 1)^2 + 2} dx$$

$$= \ln(x^2 + 2x + 3) - \frac{1}{\sqrt{2}}\tan^{-1}\left(\frac{x + 1}{\sqrt{2}}\right) + c$$

as $x^2 + 2x + 3 = (x + 1)^2 + 2 > 0$ and

recognising $\displaystyle\int \frac{f'(x)}{a^2 + \left[f(x)\right]^2} dx = \frac{1}{a}\tan^{-1}\frac{f(x)}{a} + c$,

where $f(x) = x + 1$.

Partial fractions

For the **rational function** $\dfrac{f(x)}{g(x)}$, if the degree of $f(x)$ is *greater than* the degree of $g(x)$, the numerator can be divided by the denominator and be expressed as the sum of a polynomial and another rational function, for example, $\dfrac{x^3}{x^2+1} = x - \dfrac{x}{x^2+1}$.

If the degree of $f(x)$ is *less than* the degree of $g(x)$, then the rational function can be expressed as a sum of simpler fractions called **partial fractions**, which can be integrated separately.

Example 8

Find $\displaystyle\int \dfrac{9x-2}{(2x-1)(x-3)}\,dx$.

Solution

Using partial fractions:

$$\frac{9x-2}{(2x-1)(x-3)} = \frac{A}{2x-1} + \frac{B}{x-3}$$
$$9x-2 = A(x-3) + B(2x-1)$$

We can find the values of A and B by expanding the RHS and equating coefficients or by substituting convenient values of x.

Substitute $x = 3$:

$$9(3) - 2 = A(3-3) + B(2[3]-1)$$
$$25 = 0 + 5B$$
$$B = 5$$

Substitute $x = 0$ and $B = 5$:

$$9(0) - 2 = A(0-3) + 5(2[0]-1)$$
$$-2 = -3A - 5$$
$$3 = -3A$$
$$A = -1$$

Hence,

$$\int \frac{9x-2}{(2x-1)(x-3)}\,dx = \int \frac{-1}{2x-1} + \frac{5}{x-3}\,dx$$
$$= -\frac{1}{2}\ln|2x-1| + 5\ln|x-3| + c.$$

Quadratic factors in denominator

To each irreducible quadratic factor $ax^2 + bx + c$ occurring once in the denominator of a proper function, there corresponds a single partial fraction of the form $\dfrac{Ax + B}{ax^2 + bx + c}$.

For example, $\dfrac{5x + 2}{(x - 1)(x^2 + 2x + 4)} = \dfrac{1}{x - 1} + \dfrac{-x + 2}{x^2 + 2x + 4}$.

TOPIC SUMMARY

Example 9

Find $\displaystyle\int \dfrac{3x^2 - 2x + 1}{(x^2 + 1)(x^2 + 2)}\,dx$.

Solution

Using partial fractions:

$$\frac{3x^2 - 2x + 1}{(x^2 + 1)(x^2 + 2)} = \frac{Ax + B}{x^2 + 1} + \frac{Cx + D}{x^2 + 2}$$

$$3x^2 - 2x + 1 = (Ax + B)(x^2 + 2) + (Cx + D)(x^2 + 1)$$

$$= Ax^3 + 2Ax + Bx^2 + 2B + Cx^3 + Cx + Dx^2 + D$$

$$= (A + C)x^3 + (B + D)x^2 + (2A + C)x + 2B + D$$

Equating coefficients:

$$A + C = 0 \qquad [1]$$
$$B + D = 3 \qquad [2]$$
$$2A + C = -2 \qquad [3]$$
$$2B + D = 1 \qquad [4]$$

$[3] - [1]$: $\qquad A = -2$

Substitute into $[1]$:

$$-2 + C = 0$$
$$C = 2$$

$[4] - [2]$: $\qquad B = -2$

Substitute into $[2]$:

$$-2 + D = 3$$
$$D = 5$$

Hence, $\displaystyle\int \frac{3x^2 - 2x + 1}{(x^2 + 1)(x^2 + 2)}\,dx = \int \frac{-2x - 2}{x^2 + 1} + \frac{2x + 5}{x^2 + 2}\,dx$

$$= \int \frac{-2x}{x^2 + 1} - \frac{2}{x^2 + 1} + \frac{2x}{x^2 + 2} + \frac{5}{x^2 + 2}\,dx$$

$$= -\ln(x^2 + 1) - 2\tan^{-1} x + \ln(x^2 + 2) + \frac{5}{\sqrt{2}}\tan^{-1}\left(\frac{x}{\sqrt{2}}\right) + c. \qquad (x^2 + 1 > 0)$$

Integration by parts

Integration by parts is based on the product rule.

If u and v are both functions of x, then:

$$\int u\frac{dv}{dx}\,dx = uv - \int v\frac{du}{dx}\,dx$$

or

$$\int uv'\,dx = uv - \int vu'\,dx.$$

> **Hint**
>
> **LIATE** is a handy rule advising which function should be chosen as u to differentiate.
>
> **L** Logarithmic function: $\ln x$, $\log_a x$
> **I** Inverse trigonometric functions: $\tan^{-1} x$, $\sin^{-1} x$
> **A** Algebraic functions: x^2, $2x^{10}$
> **T** Trigonometric functions: $\sin x$, $\tan x$
> **E** Exponential functions: e^x, 5^x

This formula allows us to convert a difficult integral into more manageable parts. The key is in choosing values of u and $\frac{dv}{dx}$ to split the function into. One part, u, needs to be differentiated while the other part, $\frac{dv}{dx}$, needs to be integrated.

Integration by parts allows us to integrate functions such as $\ln x$, $\tan^{-1} x$ and xe^x.

Example 10

Find $\int x\sin x\,dx$.

Solution

Let $u = x$ and $v' = \sin x$

so $u' = 1$ and $v = -\cos x$.

Using integration by parts:

$$\int uv'\,dx = uv - \int vu'\,dx$$

$$\int x\sin x\,dx = -x\cos x - \int -\cos x \times 1\,dx$$

$$= -x\cos x + \int \cos x\,dx$$

$$= -x\cos x + \sin x + c$$

Example 11

Find $\int \ln x\,dx$.

> **Hint**
>
> For integrands such as $\ln x$ and $\sin^{-1} x$ that cannot be integrated easily but which can be differentiated, let u be the integrand and $v' = 1$.

Solution

Let $u = \ln x$ and $v' = 1$

so $u' = \dfrac{1}{x}$ and $v = x$.

Using integration by parts:

$$\int uv'\,dx = uv - \int vu'\,dx$$

$$\int \ln x\,dx = (\ln x) \times x - \int \frac{1}{x} \times x\,dx$$

$$= x\ln x - \int 1\,dx$$

$$= x\ln x - x + c$$

Integration by parts can be done using the table method if $u(x)$ can be differentiated multiple times easily (and becoming simpler), while $v'(x)$ can be integrated multiple times easily.

Example 12

Find $\int x^3 \cos x\,dx$.

Solution

Let $u = x^3$ and $v' = \cos x$. Differentiate u repeatedly, integrate v' repeatedly, and alternate the 'sign':

The integral is then found by multiplying 'diagonally' pairs of terms (uv), and adding or subtracting depending on the 'sign' column.

Sign	Derivatives of u	Integrals of v'
+	x^3	$\cos x$
−	$3x^2$	$\sin x$
+	$6x$	$-\cos x$
−	6	$-\sin x$
+	0	$\cos x$

$$\int x^3 \cos x\,dx = +\,x^3(\sin x) - 3x^2(-\cos x) + 6x(-\sin x) - 6(\cos x) + c$$
$$= x^3 \sin x + 3x^2 \cos x - 6x \sin x - 6\cos x + c$$

Recurrence relations

A **recurrence relation** is a **recursive formula** that expresses an integral in terms of a similar integral with a smaller power. For example:

$$\int x^n e^x\,dx = x^n e^x - n\int x^{n-1} e^x\,dx$$

$$\int \tan^n x\,dx = \frac{\tan^{n-1}x}{n-1} - \int \tan^{n-2}x\,dx$$

With repeated application of the formula, we can eventually reduce the power of the integral to 1 or 0, when it can be easily found.

> **Hint**
> Recurrence relations usually involve integration by parts and every HSC exam has a question on it (see the HSC exam topic grid on page 124).

Example 13

Suppose that $I_n = \int x^n e^x\,dx$ for $n \geq 0$.

a Show that $I_n = x^n e^x - nI_{n-1}$.

b Hence, find I_3.

Solution

a Let $u = x^n$ and $v' = e^x$
so $u' = nx^{n-1}$ and $v = e^x$.

Using integration by parts:

$$\int uv'\,dx = uv - \int vu'\,dx$$

$$I_n = \int x^n e^x\,dx$$
$$= x^n e^x - \int e^x nx^{n-1}\,dx$$
$$= x^n e^x - n\int x^{n-1} e^x\,dx$$
$$= x^n e^x - nI_{n-1}, \text{ as required.}$$

b So $I_3 = x^3 e^x - 3I_2$

but $I_2 = x^2 e^x - 2I_1$

and $I_1 = x^1 e^x - 1I_0$.

$$I_0 = \int x^0 e^x\,dx$$
$$= \int e^x$$
$$= e^x + c$$

Hence,

$$I_3 = x^3 e^x - 3I_2$$
$$= x^3 e^x - 3(x^2 e^x - 2I_1)$$
$$= x^3 e^x - 3\left(x^2 e^x - 2[x^1 e^x - 1I_0]\right)$$
$$= x^3 e^x - 3\left(x^2 e^x - 2[xe^x - e^x]\right) + c$$
$$= x^3 e^x - 3(x^2 e^x - 2xe^x + 2e^x) + c$$
$$= e^x(x^3 - 3x^2 + 6x - 6) + c.$$

9780170459266

Example 14   ©NESA   2018 HSC EXAM, QUESTION 14(c)

Let $I_n = \int_{-3}^{0} x^n \sqrt{x + 3}\, dx$ for $n = 0, 1, 2, \ldots$

a Show that, for $n \geq 1$, $I_n = \dfrac{-6n}{3 + 2n} I_{n-1}$.

b Find the value of I_2.

Solution

a Let $u = x^n$ and $v' = (x + 3)^{\frac{1}{2}}$

$u' = nx^{n-1}$ and $v = \dfrac{2}{3}(x + 3)^{\frac{3}{2}}$.

Using integration by parts:

$$\int uv'\, dx = uv - \int vu'\, dx$$

$$I_n = \left[\frac{2}{3} x^n (x + 3)^{\frac{3}{2}} \right]_{-3}^{0} - \int_{-3}^{0} \frac{2n}{3} x^{n-1} (x + 3)^{\frac{3}{2}}\, dx$$

$$= \frac{2}{3} \left[0^n (0 + 3)^{\frac{3}{2}} - (-3)^n (-3 + 3)^{\frac{3}{2}} \right] - \frac{2n}{3} \int_{-3}^{0} x^{n-1} (x + 3)^{\frac{3}{2}}\, dx$$

$$= \frac{2}{3} [0 - 0] - \frac{2n}{3} \int_{-3}^{0} x^{n-1} (x + 3)(x + 3)^{\frac{1}{2}}\, dx$$

$$= -\frac{2n}{3} \int_{-3}^{0} x^{n-1} x \sqrt{x + 3} + x^{n-1} (3) \sqrt{x + 3}\, dx$$

$$= -\frac{2n}{3} \int_{-3}^{0} x^{n-1} \sqrt{x + 3} + 3x^{n-1} \sqrt{x + 3}\, dx$$

$$= -\frac{2n}{3} \left[I_n + 3I_{n-1} \right] \quad \text{as } I_n = \int_{-3}^{0} x^n \sqrt{x + 3}\, dx$$

$$I_n = -\frac{2n}{3} \left[I_n + 3I_{n-1} \right]$$

So $-3I_n = 2nI_n + 6nI_{n-1}$ (multiplying both sides by -3)

$-3I_n - 2nI_n = 6nI_{n-1}$

$-I_n(3 + 2n) = 6nI_{n-1}$.

$$I_n = -\frac{6n}{3 + 2n} I_{n-1}, \text{ as required.}$$

b $I_2 = -\dfrac{6(2)}{3 + 2(2)} I_1$

$= -\dfrac{12}{7} I_1$

$= -\dfrac{12}{7} \left[-\dfrac{6(1)}{3 + 2(1)} I_0 \right]$

$= -\dfrac{12}{7} \left[-\dfrac{6}{5} I_0 \right]$

$= \dfrac{72}{35} I_0$

$I_0 = \int_{-3}^{0} x^0 \sqrt{x + 3}\, dx$

$= \int_{-3}^{0} (x + 3)^{\frac{1}{2}}\, dx$

$= \left[\dfrac{2}{3} (x + 3)^{\frac{3}{2}} \right]_{-3}^{0}$

$= \left[\dfrac{2}{3} (3)^{\frac{3}{2}} \right] - \left[\dfrac{2}{3} (0)^{\frac{3}{2}} \right]$

$= \dfrac{2}{3} (\sqrt{3})^3$

$= \dfrac{2}{3} 3\sqrt{3}$

$= 2\sqrt{3}$

$I_2 = \dfrac{72}{35} I_0$

$= \dfrac{72}{35} \times 2\sqrt{3}$

$= \dfrac{144\sqrt{3}}{35}$

Practice set 1

Multiple-choice questions

Solutions start on page 116.

Question 1

Which substitution would be best for finding $\int 6x^2(2x^3 - 1)^4 \, dx$?

A $u = 2x^3$ **B** $u = 2x^3 - 1$ **C** $u = (2x^3 - 1)^4$ **D** $u = 6x^2$

Question 2

Find $\int \sin x \cos^4 x \, dx$.

A $\dfrac{1}{5}\cos^5 x + c$ **B** $-\dfrac{1}{5}\cos^5 x + c$ **C** $\dfrac{1}{5}\sin^5 x + c$ **D** $-\dfrac{1}{5}\sin^5 x + c$

Question 3

Find $\int -x\cos x \, dx$.

A $-x\sin x + \cos x + c$ **B** $-x\sin x - \cos x + c$ **C** $x\sin x + \cos x + c$ **D** $x\sin x - \cos x + c$

Question 4

Find $\int \tan^{-1} x \, dx$.

A $x\tan^{-1} x + \dfrac{1}{2}\ln(1 + x^2) + c$ **B** $x\tan^{-1} x + 2\ln(1 + x^2) + c$

C $x\tan^{-1} x - \dfrac{1}{2}\ln(1 + x^2) + c$ **D** $x\tan^{-1} x - 2\ln(1 + x^2) + c$

Question 5

What is a primitive of $\dfrac{\cos x}{\sin^3 x}$?

A $\dfrac{1}{2}\operatorname{cosec}^2 x$ **B** $-\dfrac{1}{2}\operatorname{cosec}^2 x$ **C** $\dfrac{1}{4}\operatorname{cosec}^4 x$ **D** $-\dfrac{1}{4}\operatorname{cosec}^4 x$

Question 6

What is $\int \dfrac{1}{\sqrt{1 - 9x^2}} \, dx$?

A $\dfrac{1}{3}\sin^{-1}\left(\dfrac{x}{3}\right) + c$ **B** $\dfrac{1}{3}\sin^{-1} 3x + c$ **C** $\sin^{-1}\left(\dfrac{x}{3}\right) + c$ **D** $\sin^{-1} 3x + c$

Question 7

Evaluate $\int_0^1 \dfrac{e^{2x}}{e^{2x} + 1} \, dx$.

A $\ln(e + 1)$ **B** $\dfrac{1}{2}\ln(e^2 + 1)$ **C** $\dfrac{1}{2}\ln\left(\dfrac{e^2 + 1}{2}\right)$ **D** $\ln(e^2 + 1)$

Question 8 ⬤⬤○

If $\dfrac{6}{(2x+1)(1-x)} = \dfrac{A}{2x+1} + \dfrac{B}{1-x}$, find A and B.

A $A = 1, B = 2$ **B** $A = -1, B = 2$ **C** $A = 4, B = 2$ **D** $A = 4, B = -2$

Question 9 ⬤⬤○

What is $\displaystyle\int \dfrac{dx}{x^2 + 4x + 13}$?

A $\dfrac{1}{3}\tan^{-1}\left(\dfrac{x+2}{3}\right) + c$ **B** $\tan^{-1}\left(\dfrac{x+2}{3}\right) + c$ **C** $\dfrac{1}{6}\ln\left[\dfrac{x-1}{x+5}\right] + c$ **D** $\ln\left[\dfrac{x-1}{x+5}\right] + c$

Question 10 ⬤⬤○

Which expression is equal to $\displaystyle\int x^2 \cos x\, dx$?

A $x^2 \sin x - \displaystyle\int 2x \sin x\, dx$ **B** $2x \sin x + \displaystyle\int x^2 \sin x\, dx$

C $x^2 \sin x + \displaystyle\int 2x \sin x\, dx$ **D** $2x \sin x - \displaystyle\int x^2 \sin x\, dx$

Question 11 ⬤⬤○

Find $\displaystyle\int \dfrac{1}{\sqrt{-x^2 + 4x + 5}}\, dx$.

A $\sin^{-1}\left(\dfrac{x-2}{3}\right) + c$ **B** $\cos^{-1}\left(\dfrac{x-2}{3}\right) + c$ **C** $\dfrac{1}{3}\sin^{-1}\left(\dfrac{x-2}{3}\right) + c$ **D** $\dfrac{1}{3}\cos^{-1}\left(\dfrac{x-2}{3}\right) + c$

Question 12 ©NESA | 2013 HSC EXAM, QUESTION 1 MODIFIED ⬤○○

Which expression is equal to $\displaystyle\int \tan x\, dx$?

A $\sec^2 x + c$ **B** $-\ln|\cos x| + c$

C $\dfrac{\tan^2 x}{2} + c$ **D** $\ln|\sec x + \tan x| + c$

Question 13 ©NESA | 2014 HSC EXAM, QUESTION 7 ⬤⬤⬤

Which expression is equal to $\displaystyle\int \dfrac{1}{1 - \sin x}\, dx$?

A $\tan x - \sec x + c$ **B** $\tan x + \sec x + c$

C $\log_e(1 - \sin x) + c$ **D** $\dfrac{\log_e(1 - \sin x)}{-\cos x} + c$

Question 14 ⬤⬤⬤

Which expression is equal to $\displaystyle\int \dfrac{dx}{\sqrt{12 + 4x - x^2}}$?

A $\sin^{-1}\left(\dfrac{x-2}{2\sqrt{3}}\right) + c$ **B** $\sin^{-1}\left(\dfrac{x-2}{4}\right) + c$ **C** $\sin^{-1}\left(\dfrac{x+2}{2\sqrt{3}}\right) + c$ **D** $\sin^{-1}\left(\dfrac{x+2}{4}\right) + c$

Question 15 ⬤⬤○

Evaluate $\displaystyle\int_0^2 |1 - 2x|\, dx$.

A 0   **B** 1   **C** 2.5   **D** 3

Question 16 ●●●

Which definite integral has the smallest value?

A $\int_0^{\frac{\pi}{2}} \sin x \, dx$

B $\int_0^{\frac{\pi}{2}} \sin^2 x \, dx$

C $\int_0^{\frac{\pi}{2}} (1 - \sin x) \, dx$

D $\int_0^{\frac{\pi}{2}} (1 - \sin^2 x) \, dx$

Question 17 ●●●

If n is a non-negative integer, for what value(s) of n is $\int_0^1 (1 - x)^n \, dx = \int_0^1 (1 + x)^n \, dx$?

A There is no solution

B $n = 0$

C n is odd

D All values of n

Question 18 ●●●

Which equation is true for $I_n = \int_0^1 x^n e^x \, dx$?

A $I_n = e - nI_{n-1}$

B $I_n = -nI_{n-1}$

C $I_n = e + nI_{n-1}$

D $I_n = nI_{n-1}$

Question 19 ©NESA 2020 HSC EXAM, QUESTION 10 ●●●

Which of the following is equal to $\int_0^{2a} f(x) \, dx$?

A $\int_0^a f(x) - f(2a - x) \, dx$

B $\int_0^a f(x) + f(2a - x) \, dx$

C $2\int_0^a f(x - a) \, dx$

D $\int_0^a \frac{1}{2} f(2x) \, dx$

Question 20 ©NESA 2012 HSC EXAM, QUESTION 10 ●●●

Without evaluating the integrals, which one of the following integrals is greater than zero?

A $\int_{-\frac{\pi}{2}}^{\frac{\pi}{2}} \frac{x}{2 + \cos x} \, dx$

B $\int_{-\pi}^{\pi} x^3 \sin x \, dx$

C $\int_{-1}^1 \left(e^{-x^2} - 1\right) dx$

D $\int_{-2}^2 \tan^{-1}(x^3) \, dx$

Practice set 2

Short-answer questions

Solutions start on page 118.

Question 1 (2 marks) ●●○

Find $\int x \ln x \, dx$.

2 marks

Question 2 (2 marks) ○●○

Find $\int x e^x dx$.

2 marks

Question 3 (3 marks) ●●○

Find $\int \dfrac{\ln x}{x} \, dx$.

3 marks

Question 4 (3 marks) ●●○

By completing the square, find $\int \dfrac{dx}{x^2 + 6x + 17}$.

3 marks

Question 5 (3 marks) ●●○

Evaluate $\int_1^{e^2} x \ln x \, dx$.

3 marks

Question 6 (2 marks) ●●○

Evaluate $\int_0^{\ln 2} x e^x \, dx$.

2 marks

Question 7 (2 marks) ●●○

Find $\int \dfrac{8}{x^2 - 4} \, dx$.

2 marks

Question 8 (2 marks) ●●○

Evaluate $\int_1^5 x\sqrt{x - 1} \, dx$.

2 marks

Question 9 (3 marks) ●●○

Evaluate $\int_1^2 \dfrac{e^{2x}}{e^{2x} - 1} \, dx$.

3 marks

Question 10 (3 marks) ●●○

Find $\int \sin^3 x \, dx$.

3 marks

Question 11 (3 marks) ●●○

Find $\int_1^e \dfrac{1}{x(1 + \ln x)} \, dx$.

3 marks

Question 12 (3 marks) ●●○

Find $\int \dfrac{x^2 + 6x}{x^2 + 6x + 10} \, dx$.

3 marks

Question 13 (3 marks) ●●●

Find $\int_0^{\frac{1}{2}} \sin^{-1} 2x \, dx$. 3 marks

Question 14 (3 marks) ●●●

Find $\int xe^{-x} \, dx$. 3 marks

Question 15 (4 marks) ●●●

a Find real numbers A, B and C such that 2 marks

$$\frac{2}{(x^2+1)(x-1)} = \frac{Ax+B}{x^2+1} + \frac{C}{x-1}.$$

b Hence, find $\int \frac{2}{(x^2+1)(x-1)} \, dx$. 2 marks

Question 16 (3 marks) ●●●

Using the substitution $x = \cos^2 \theta$, or otherwise, evaluate $\int_{\frac{1}{2}}^1 \sqrt{\frac{x}{1-x}} \, dx$. 3 marks

Question 17 (4 marks) ●●●

a Use a suitable substitution to show that $\int_0^a f(x) \, dx = \int_0^a f(a-x) \, dx$. 1 mark

b Hence, or otherwise, show that $\int_0^{\frac{\pi}{4}} \frac{\cos x}{\cos x + \sin x} \, dx = \frac{\pi}{8} + \frac{1}{4} \ln 2$. 3 marks

Question 18 (5 marks) ●●●

a Differentiate $\cos^{n-1} \theta \sin \theta$, expressing the result in terms of $\cos \theta$ only. 2 marks

b Hence, or otherwise, deduce that 2 marks

$$\int_0^{\frac{\pi}{2}} \cos^n \theta \, d\theta = \frac{n-1}{n} \int_0^{\frac{\pi}{2}} \cos^{n-2} \theta \, d\theta$$

for $n > 1$.

c Find $\int_0^{\frac{\pi}{2}} \cos^4 \theta \, d\theta$. 1 mark

Question 19 (4 marks) ©NESA 2019 HSC EXAM, QUESTION 15(a) ●●●

a Show that $\int_{-a}^a \frac{f(x)}{f(x)+f(-x)} \, dx = \int_{-a}^a \frac{f(-x)}{f(x)+f(-x)} \, dx$. 2 marks

b Hence, or otherwise, evaluate $\int_{-1}^1 \frac{e^x}{e^x + e^{-x}} \, dx$. 2 marks

Question 20 (5 marks) ©NESA 2017 HSC EXAM, QUESTION 15(a) ●●●

Let $I_n = \int_0^1 x^n \sqrt{1-x^2} \, dx$, for $n = 0, 1, 2, \dots$

a Find the value of I_1. 1 mark

b Using integration by parts, or otherwise, show that for $n \geq 2$ 3 marks

$$I_n = \left(\frac{n-1}{n+2}\right) I_{n-2}.$$

c Find the value of I_5. 1 mark

Practice set 1

Worked solutions

1 B

$\int 6x^2(2x^3 - 1)^4 \, dx$, since $\frac{d}{dx}(2x^3 - 1) = 6x^2$

then this would make the best substitution.

2 B

Let $u = \cos x$.

$$du = -\sin x \, dx$$
$$dx = -\frac{du}{\sin x}$$
$$\int \sin x \cos^4 x \, dx = -\int u^4 \, du$$
$$= -\frac{1}{5}u^5 + c$$
$$= -\frac{1}{5}\cos^5 x + c$$

3 B

Using integration by parts:

$u = -x, \ v' = \cos x$
$u' = -1, \ v = \sin x$

$$\int -x\cos x \, dx = -x\sin x + \int \sin x \, dx$$

4 C

$u = \tan^{-1} x \qquad v' = 1$
$u' = \dfrac{1}{1 + x^2} \qquad v = x$

$$\int \tan x \, dx = x\tan^{-1} x - \int \frac{x}{1 + x^2} \, dx$$
$$= x\tan^{-1} x - \frac{1}{2}\ln\left(1 + x^2\right) + c$$

5 B

Let $u = \sin x, \ du = \cos x \, dx$:

$$\int \frac{\cos x}{\sin^3 x} \, dx = \int u^{-3} \, du$$
$$= -\frac{1}{2}u^{-2} + c$$
$$= -\frac{1}{2}\frac{1}{\sin^2 x} + c$$
$$= -\frac{1}{2}\operatorname{cosec}^2 x + c$$

6 B

$$\int \frac{1}{\sqrt{1 - 9x^2}} \, dx = \frac{1}{3}\int \frac{3}{\sqrt{1 - (3x)^2}} \, dx$$
$$= \frac{1}{3}\sin^{-1} 3x + c$$

7 C

$$\int_0^1 \frac{e^{2x}}{e^{2x} + 1} \, dx = \frac{1}{2}\int_0^1 \frac{2e^{2x}}{e^{2x} + 1} \, dx$$
$$= \frac{1}{2}\left[\ln(e^{2x} + 1)\right]_0^1 \quad e^{2x} + 1 > 0$$
$$= \frac{1}{2}\left[\ln(e^2 + 1) - \ln(e^0 + 1)\right]$$
$$= \frac{1}{2}\left[\ln(e^2 + 1) - \ln 2\right]$$
$$= \frac{1}{2}\ln\left(\frac{e^2 + 1}{2}\right)$$

8 C

$$\frac{6}{(2x + 1)(1 - x)} = \frac{A}{2x + 1} + \frac{B}{1 - x}$$
$$\therefore 6 = A(1 - x) + B(2x + 1)$$
$$6 = A - Ax + 2Bx + B$$
$$= (-A + 2B)x + A + B$$

Equating coefficients:

$-A + 2B = 0$ and $A + B = 6$

Solving simultaneously by adding the equations:

$3B = 6$
$B = 2$

Substitute into 2nd equation:

$A + 2 = 6$
$A = 4$

9 A

$$\int \frac{dx}{x^2 + 4x + 13} = \int \frac{dx}{(x^2 + 4x + 4) + 9}$$
$$= \int \frac{dx}{(x + 2)^2 + 3^2}$$
$$= \frac{1}{3}\tan^{-1}\left(\frac{x + 2}{3}\right) + c$$

10 A

Using integration by parts

$u = x^2, \ v' = \cos x$

$u' = 2x, \ v = \sin x$

$$\int x^2 \cos x \, dx = x^2 \sin x - \int 2x\sin x \, dx$$

11 A

$$\int \frac{1}{\sqrt{-x^2 + 4x + 5}} \, dx = \int \frac{1}{\sqrt{-(x^2 - 4x + 4) + 9}} \, dx$$

$$= \int \frac{1}{\sqrt{3^2 - (x - 2)^2}} \, dx$$

$$= \sin^{-1}\left(\frac{x - 2}{3}\right) + c$$

12 B

$$\int \tan x \, dx = \int \frac{\sin x}{\cos x} \, dx$$

$$= -\int \frac{-\sin x}{\cos x} \, dx$$

$$= -\ln|\cos x| + c$$

13 B

$$\int \frac{1}{1 - \sin x} \, dx = \int \frac{1}{1 - \sin x} \times \frac{1 + \sin x}{1 + \sin x} \, dx$$

$$= \int \frac{1 + \sin x}{1 - \sin^2 x} \, dx$$

$$= \int \frac{1 + \sin x}{\cos^2 x} \, dx$$

$$= \int \frac{1}{\cos^2 x} + \frac{\sin x}{\cos^2 x} \, dx$$

$$= \int \sec^2 x + \sec x \tan x \, dx$$

$$= \tan x + \sec x + c$$

14 B

$$\int \frac{dx}{\sqrt{12 + 4x - x^2}}$$

$$= \int \frac{dx}{\sqrt{16 - (x^2 - 4x + 4)}}$$

$$= \int \frac{dx}{\sqrt{4^2 - (x - 2)^2}}$$

$$= \sin^{-1}\left(\frac{x - 2}{4}\right) + c$$

15 C

$$\int_0^2 |1 - 2x| \, dx$$

$$= \int_0^{\frac{1}{2}} 1 - 2x \, dx + \int_{\frac{1}{2}}^2 2x - 1 \, dx$$

$$= \left[x - x^2\right]_0^{\frac{1}{2}} + \left[x^2 - x\right]_{\frac{1}{2}}^2$$

$$= \left[\frac{1}{2} - \left(\frac{1}{2}\right)^2 - 0\right] + \left[2^2 - 2 - \left\{\left(\frac{1}{2}\right)^2 - \frac{1}{2}\right\}\right]$$

$$= \left(\frac{1}{4}\right) + \left(2 + \frac{1}{4}\right)$$

$$= 2\frac{1}{2}$$

16 C

By observing areas when graphs of the 4 functions are compared or

A: $\int_0^{\frac{\pi}{2}} \sin x \, dx = 1$

B: $\int_0^{\frac{\pi}{2}} \sin^2 x \, dx = \frac{\pi}{4}$

C: $\int_0^{\frac{\pi}{2}} (1 - \sin x) \, dx = \frac{\pi}{2} - 1$

D: $\int_0^{\frac{\pi}{2}} (1 - \sin^2 x) \, dx = \frac{\pi}{4}$

17 B

LHS $= \int_0^1 (1 - x)^n \, dx = \frac{1}{n + 1}$

RHS $= \int_0^1 (1 + x)^n \, dx = \frac{2^{n+1} - 1}{n + 1}$

Equating and solving;

$$1 = 2^{n+1} - 1$$

$$2^{n+1} = 2$$

$$\therefore n + 1 = 1$$

$$\text{So } n = 0.$$

18 A

$\int_0^1 x^n e^x \, dx$, using integration by parts, putting $u = x^n$, $v' = e^x$.

$$\int_0^1 x^n e^x \, dx = \left[x^n e^x\right]_0^1 - n \int_0^1 x^{n-1} e^x \, dx$$

$$= e - n I_{n-1}$$

19 B

$$\int_0^a f(x) + f(2a - x) \, dx$$

$$= \int_0^a f(x) \, dx + \int_0^a f(2a - x) \, dx$$

But if $u = 2a - x$

$$du = -dx.$$

When $x = 0$, $u = 2a$.

When $x = a$, $u = 2a - a = a$.

$$\int_0^a f(2a - x) \, dx = \int_{2a}^a -f(u) \, du$$

$$= \int_a^{2a} f(u) \, du$$

$$= \int_a^{2a} f(x) \, dx$$

Hence,

$$\int_0^a f(x) + f(2a - x) \, dx$$

$$= \int_0^a f(x) \, dx + \int_a^{2a} f(x) \, dx$$

$$= \int_0^{2a} f(x) \, dx.$$

9780170459266

WORKED SOLUTIONS

20 B

Consider the graphs of the functions on both sides of the y-axis as the limits of integration are symmetrical.

A $y = \dfrac{x}{2 + \cos x}$: between $-\dfrac{\pi}{2}$ and $\dfrac{\pi}{2}$,

cos x is positive and symmetrical, so cos $x + 1$ is positive and symmetrical (an even function). In the numerator, x is an odd function, so the whole function is odd. Therefore, the integral = 0.

B $y = x^3 \sin x$: between $-\pi$ and π, x^3 is odd and sin x is odd, so the whole function is even and above the x-axis (except at $x = 0$). The integral > 0.

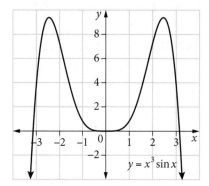

C $y = e^{-x^2} - 1$: between -1 and 1, e^{-x^2} is positive and symmetrical (an even function), but its value is between e^{-1} and e^0, that is, 0.37 and 1, so $e^{-x^2} - 1$ is negative and below the y-axis (except at $x = 0$), so the integral < 0.

D $y = \tan^{-1}(x^3)$: between -2 and 2, x^3 is odd, and its value is between $(-2)^3$ and 2^3, that is, -8 and 8. The inverse tan function is also odd, its value between $\tan^{-1}(-8)$ and $\tan^{-1}(8)$, so the integral = 0.

So option B is the only integral that is greater than 0.

Practice set 2

Worked solutions

Question 1

Integrating by parts:

$$u = \ln x \qquad v' = x$$
$$u' = \frac{1}{x} \qquad v = \frac{1}{2}x^2$$

$$\int x \ln x \, dx = \frac{1}{2}x^2 \ln x - \int \frac{1}{2x}x^2 dx$$
$$= \frac{1}{2}x^2 \ln x - \int \frac{1}{2}x \, dx$$
$$= \frac{1}{2}x^2 \ln x - \frac{1}{4}x^2 + c$$

Question 2

Integrating by parts:

$$u = x \qquad v' = e^x$$
$$u' = 1 \qquad v = e^x$$

$$\int xe^x \, dx = xe^x - \int e^x \, dx$$
$$= xe^x - e^x + c$$

Question 3

Let $u = \ln x$.

$$du = \frac{1}{x}dx$$
$$dx = x \, du$$

$$\int \frac{\ln x}{x}dx = \int \frac{u}{x} x \, du$$
$$= \int u \, du$$
$$= \frac{1}{2}u^2 + c$$
$$= \frac{1}{2}(\ln x)^2 + c$$

Question 4

$$\int \frac{dx}{x^2 + 6x + 17} = \int \frac{dx}{(x^2 + 6x + 9) + 8}$$

$$= \int \frac{dx}{(x + 3)^2 + 8}$$

$$= \frac{1}{2\sqrt{2}} \tan^{-1}\left(\frac{x + 3}{2\sqrt{2}}\right) + c$$

Question 5

From Question **1**, after integrating by parts:

$$\int x \ln x \, dx = \frac{1}{2}x^2 \ln x - \frac{1}{4}x^2 + c$$

$$\int_1^{e^2} x \ln x \, dx = \left[\frac{1}{2}x^2 \ln x - \frac{1}{4}x^2\right]_1^{e^2}$$

$$= \left(\frac{1}{2}e^4 \ln e^2 - \frac{1}{4}e^4\right) - \left(0 - \frac{1}{4}\right)$$

$$= \left(\frac{1}{2}e^4 (2) - \frac{1}{4}e^4\right) + \frac{1}{4}$$

$$= \frac{3}{4}e^4 + \frac{1}{4}$$

Question 6

From Question **2**, after integrating by parts:

$$\int xe^x \, dx = xe^x - e^x + c$$

$$\int_0^{\ln 2} xe^x \, dx = \left[xe^x - e^x\right]_0^{\ln 2}$$

$$= [\ln 2(2) - 2] - [0 - 1]$$

$$= 2\ln 2 - 2 + 1$$

$$= 2\ln 2 - 1 \text{ (or } \ln 4 - 1)$$

Question 7

Let $\dfrac{8}{x^2 - 4} = \dfrac{A}{x - 2} + \dfrac{B}{x + 2}$.

$$8 = A(x + 2) + B(x - 2)$$
$$8 = Ax + 2A + Bx - 2B$$
$$= (A + B)x + 2A - 2B$$

$A + B = 0$	$2A - 2B = 8$
$A = -B$	$2(-B) - 2B = 8$
$A = 2$	$B = -2$

$$\int \frac{8}{x^2 - 4} dx = \int \frac{2}{x - 2} - \frac{2}{x + 2} dx$$

$$= 2\ln|x - 2| - 2\ln|x + 2| + c$$

$$= 2\ln\left|\frac{x - 2}{x + 2}\right| + c$$

Question 8

Let $u = x - 1$ When $x = 1$, $u = 0$.
    $x = u + 1$ When $x = 5$, $u = 4$.
    $du = dx$

$$\int_1^5 x\sqrt{x - 1} \, dx = \int_0^4 (u + 1)\sqrt{u} \, du$$

$$= \int_0^4 u^{\frac{3}{2}} + u^{\frac{1}{2}} du$$

$$= \left[\frac{2}{5}u^{\frac{5}{2}} + \frac{2}{3}u^{\frac{3}{2}}\right]_0^4$$

$$= \left(\frac{2}{5}4^{\frac{5}{2}} + \frac{2}{3}4^{\frac{3}{2}}\right) - \left(\frac{2}{5}0^{\frac{5}{2}} + \frac{2}{3}0^{\frac{3}{2}}\right)$$

$$= \frac{2}{5}(32) + \frac{2}{3}(8) - 0$$

$$= \frac{272}{15}$$

Question 9

$$\int_1^2 \frac{e^{2x}}{e^{2x} - 1} dx = \frac{1}{2}\int_1^2 \frac{2e^{2x}}{e^{2x} - 1} dx$$

$$= \frac{1}{2}\left[\left|\ln(e^{2x} - 1)\right|\right]_1^2$$

$$= \frac{1}{2}\left[\ln(e^4 - 1) - \ln(e^2 - 1)\right]$$

$$= \frac{1}{2}\ln\left(\frac{e^4 - 1}{e^2 - 1}\right)$$

$$= \frac{1}{2}\ln\left(\frac{[e^2 + 1][e^2 - 1]}{e^2 - 1}\right)$$

$$= \frac{1}{2}\ln(e^2 + 1)$$

Question 10

$$\int \sin^3 x \, dx = \int \sin x \, (\sin^2 x) \, dx$$

$$= \int \sin x \, (1 - \cos^2 x) \, dx$$

$$= \int \sin x - \sin x \, \cos^2 x \, dx$$

$$= -\cos x + \frac{1}{3}\cos^3 x + c$$

$$= \frac{1}{3}\cos^3 x - \cos x + c$$

Question 11

$$\int_1^e \frac{1}{x(1 + \ln x)} dx = \int_1^e \frac{\frac{1}{x}}{1 + \ln x} dx$$

$$= \left[\ln|1 + \ln x|\right]_1^e$$

$$= \ln(1 + \ln e) - \ln(1 + \ln 1)$$

$$= \ln(1 + 1) - \ln(1 + 0)$$

$$= \ln 2$$

Question 12

$$\int \frac{x^2 + 6x}{x^2 + 6x + 10}\,dx = \int \frac{(x^2 + 6x + 10) - 10}{x^2 + 6x + 10}\,dx$$

$$= \int 1 - \frac{10}{x^2 + 6x + 9 + 1}\,dx$$

$$= \int 1 - \frac{10}{(x + 3)^2 + 1}\,dx$$

$$= x - 10\tan^{-1}(x + 3) + c$$

Question 13

Integrating by parts:

$$u = \sin^{-1} 2x \qquad\qquad v' = 1$$

$$u' = \frac{2}{\sqrt{1 - 4x^2}} \qquad\qquad v = x$$

$$\int_0^{\frac{1}{2}} \sin^{-1} 2x\,dx = \left[x\sin^{-1}2x\right]_0^{\frac{1}{2}} - \int_0^{\frac{1}{2}} \frac{2x}{\sqrt{1 - 4x^2}}\,dx$$

$$= \left[\frac{1}{2}\sin^{-1}1 - 0\right] + \frac{1}{2}\left[\sqrt{1 - 4x^2}\right]_0^{\frac{1}{2}}$$

$$= \left[\frac{1}{2}\left(\frac{\pi}{2}\right)\right] + \frac{1}{2}\left[\sqrt{1 - 1} - \sqrt{1 - 0}\right]$$

$$= \frac{\pi}{4} + \frac{1}{2}(0 - 1)$$

$$= \frac{\pi}{4} - \frac{1}{2}$$

Question 14

Integrating by parts:

$$u = x \qquad\qquad v' = e^{-x}$$

$$u' = 1 \qquad\qquad v = -e^{-x}$$

$$\int xe^{-x}\,dx = -xe^{-x} - \int -e^{-x}\,dx$$

$$= -xe^{-x} - e^{-x} + c \quad \text{or} \quad -e^{-x}(x + 1) + c$$

Question 15

a $\dfrac{2}{(x^2 + 1)(x - 1)} = \dfrac{Ax + B}{x^2 + 1} + \dfrac{C}{x - 1}$

So $2 = (Ax + B)(x - 1) + C(x^2 + 1)$.

Equating coefficients:

$$2 = Ax^2 - Ax + Bx - B + Cx^2 + C$$

$$= (A + C)x^2 + (-A + B)x - B + C$$

$$A + C = 0 \qquad -A + B = 0 \qquad -B + C = 2$$

$$C = -A \qquad\quad B = A \qquad -A + (-A) = 2$$

$$C = 1 \qquad\qquad B = -1 \qquad\quad -2A = 2$$

$$A = -1$$

b $\displaystyle\int \frac{2}{(x^2 + 1)(x - 1)}\,dx$

$$= \int \frac{-x - 1}{x^2 + 1} + \frac{1}{x - 1}\,dx$$

$$= \int -\frac{1}{2}\left(\frac{2x}{x^2 + 1}\right) - \frac{1}{x^2 + 1} + \frac{1}{x - 1}\,dx$$

$$= -\frac{1}{2}\ln\left(x^2 + 1\right) - \tan^{-1}x + \ln|x - 1| + c$$

Question 16

$$x = \cos^2\theta,$$

$$dx = -2\cos\theta\sin\theta\,d\theta$$

When $x = \dfrac{1}{2}$:

$$\frac{1}{2} = \cos^2\theta$$

$$\cos\theta = \frac{1}{\sqrt{2}}$$

$$\theta = \frac{\pi}{4}$$

When $x = 1$,

$$1 = \cos^2\theta$$

$$\cos\theta = 1$$

$$\theta = 0$$

$$\int_{\frac{1}{2}}^{1} \sqrt{\frac{x}{1 - x}}\,dx = \int_{\frac{\pi}{4}}^{0} \sqrt{\frac{\cos^2\theta}{1 - \cos^2\theta}}\,(-2\cos\theta\sin\theta)\,d\theta$$

$$= -2\int_{\frac{\pi}{4}}^{0} \sqrt{\frac{\cos^2\theta}{\sin^2\theta}}\,\cos\theta\sin\theta\,d\theta$$

$$= -2\int_{\frac{\pi}{4}}^{0} \frac{\cos\theta}{\sin\theta}\,\cos\theta\sin\theta\,d\theta$$

$$= 2\int_0^{\frac{\pi}{4}} \cos^2\theta\,d\theta$$

$$= 2\int_0^{\frac{\pi}{4}} \frac{1}{2}(1 + \cos 2\theta)\,d\theta$$

$$= \int_0^{\frac{\pi}{4}} (1 + \cos 2\theta)\,d\theta$$

$$= \left[\theta + \frac{1}{2}\sin 2\theta\right]_0^{\frac{\pi}{4}}$$

$$= \left[\frac{\pi}{4} + \frac{1}{2}\sin\left(\frac{\pi}{2}\right)\right] - \left[0 + \frac{1}{2}\sin 0\right]$$

$$= \left[\frac{\pi}{4} + \frac{1}{2}(1)\right] - 0$$

$$= \frac{\pi + 2}{4}$$

Question 17

a $\int_0^a f(x)\,dx$, putting $x = a - u$, $dx = -du$

Let $u = a - x$:

$du = -dx$

$dx = -du$

$x = a - u$

So $\int_0^a f(x)\,dx = \int_a^0 f(a - u)(-du)$

$\qquad = -\int_a^0 f(a - u)\,du$

$\qquad = \int_0^a f(a - u)\,du$

$\qquad = \int_0^a f(a - x)\,dx$, as required.

b $\int_0^{\frac{\pi}{4}} \dfrac{\cos x}{\cos x + \sin x}\,dx$, so let $a = \dfrac{\pi}{4}$ and using the result from part **a**:

$$\int_0^{\frac{\pi}{4}} \frac{\cos x}{\cos x + \sin x}\,dx = \int_0^{\frac{\pi}{4}} \frac{\cos\left(\dfrac{\pi}{4} - x\right)}{\cos\left(\dfrac{\pi}{4} - x\right) + \sin\left(\dfrac{\pi}{4} - x\right)}\,dx$$

$$= \int_0^{\frac{\pi}{4}} \frac{\cos\dfrac{\pi}{4}\cos x + \sin\dfrac{\pi}{4}\sin x}{\cos\dfrac{\pi}{4}\cos x + \sin\dfrac{\pi}{4}\sin x + \sin\dfrac{\pi}{4}\cos x - \cos\dfrac{\pi}{4}\sin x}\,dx$$

$$= \int_0^{\frac{\pi}{4}} \frac{\cos x + \sin x}{\cos x + \sin x + \cos x - \sin x}\,dx \quad \left(\text{as } \cos\frac{\pi}{4} = \sin\frac{\pi}{4} = \frac{1}{\sqrt{2}}\right)$$

$$= \int_0^{\frac{\pi}{4}} \frac{\cos x + \sin x}{2\cos x}\,dx$$

$$= \frac{1}{2}\int_0^{\frac{\pi}{4}} 1 + \frac{\sin x}{\cos x}\,dx$$

$$= \frac{1}{2}\Big[x - \ln|\cos x|\Big]_0^{\frac{\pi}{4}}$$

$$= \frac{1}{2}\left[\frac{\pi}{4} - \ln\left|\cos\frac{\pi}{4}\right| - \left(0 - \ln|\cos 0|\right)\right]$$

$$= \frac{1}{2}\left[\frac{\pi}{4} - \ln\left(\frac{1}{\sqrt{2}}\right) - 0\right]$$

$$= \frac{1}{2}\left(\frac{\pi}{4} - \ln 2^{-\frac{1}{2}}\right)$$

$$= \frac{1}{2}\left(\frac{\pi}{4} + \frac{1}{2}\ln 2\right)$$

$$= \frac{\pi}{8} + \frac{1}{4}\ln 2$$

Question 18

a By the product rule:

$$\frac{d}{d\theta}(\cos^{n-1}\theta\sin\theta)$$

$$= (n-1)\cos^{n-2}\theta \times (-\sin\theta) \times \sin\theta + \cos^{n-1}\theta \times \cos\theta$$

$$= (n-1)\cos^{n-2}\theta(-\sin^2\theta) + \cos^n\theta$$

$$= (n-1)\cos^{n-2}\theta(\cos^2\theta - 1) + \cos^n\theta$$

$$= (n-1)\cos^n\theta - (n-1)\cos^{n-2}\theta + \cos^n\theta$$

$$= n\cos^n\theta - (n-1)\cos^{n-2}\theta$$

b From part **a**:

$$\int_0^{\frac{\pi}{2}} n\cos^n\theta - (n-1)\cos^{n-2}\theta\, d\theta$$

$$= \left[\cos^{n-1}\theta\sin\theta\right]_0^{\frac{\pi}{2}}$$

$$= \left[\cos^{n-1}\left(\frac{\pi}{2}\right)\sin\left(\frac{\pi}{2}\right)\right] - \left[\cos^{n-1}0\sin0\right]$$

$$= 0 \times 1 - 1 \times 0$$

$$= 0$$

$$\int_0^{\frac{\pi}{2}} n\cos^n\theta\, d\theta = \int_0^{\frac{\pi}{2}}(n-1)\cos^{n-2}\theta\, d\theta$$

$$\int_0^{\frac{\pi}{2}}\cos^n\theta\, d\theta = \frac{n-1}{n}\int_0^{\frac{\pi}{2}}\cos^{n-2}\theta\, d\theta$$

c From part **b**:

$$\int_0^{\frac{\pi}{2}}\cos^4\theta\, d\theta = \frac{3}{4}\int_0^{\frac{\pi}{2}}\cos^2\theta\, d\theta$$

$$= \frac{3}{4}\int_0^{\frac{\pi}{2}}\frac{1}{2}(1+\cos 2\theta)\, d\theta$$

$$= \frac{3}{4}\left[\frac{1}{2}\left(\theta + \frac{1}{2}\sin 2\theta\right)\right]_0^{\frac{\pi}{2}}$$

$$= \frac{3}{8}\left(\left[\frac{\pi}{2} + \frac{1}{2}\sin\pi\right] - \left[0 + \frac{1}{2}\sin 0\right]\right)$$

$$= \frac{3}{8}\left(\frac{\pi}{2} + 0 - 0\right)$$

$$= \frac{3\pi}{16}$$

Question 19

a Let $u = -x$

$$du = -dx$$

When $\quad x = -a, \quad u = a$

$$\qquad\quad x = a, \quad\; u = -a.$$

$$\int_{-a}^{a}\frac{f(x)}{f(x) + f(-x)}\, dx$$

$$= \int_{a}^{-a}\frac{f(-u)}{f(-u) + f(u)}(-du)$$

$$= -\int_{a}^{-a}\frac{f(-u)}{f(-u) + f(u)}\, du$$

$$= \int_{-a}^{a}\frac{f(-u)}{f(-u) + f(u)}\, du$$

$$= \int_{-a}^{a}\frac{f(-x)}{f(-x) + f(x)}\, dx, \text{ as required.}$$

b Let $I = \int_{-1}^{1}\frac{e^x}{e^x + e^{-x}}\, dx = \int_{-1}^{1}\frac{e^{-x}}{e^{-x} + e^{x}}\, dx$,

using result from part **a**.

Therefore, $2I = \int_{-1}^{1}\frac{e^x}{e^x + e^{-x}}\, dx + \int_{-1}^{1}\frac{e^{-x}}{e^{-x} + e^{x}}\, dx$

$$= \int_{-1}^{1}\frac{e^{-x} + e^{x}}{e^{x} + e^{-x}}\, dx$$

$$= \int_{-1}^{1}1\, dx$$

$$= [x]_{-1}^{1}$$

$$= 1 - (-1)$$

$$= 2.$$

So $I = \int_{-1}^{1}\frac{e^x}{e^x + e^{-x}}\, dx = 1$.

Question 20

a $I_1 = \int_0^1 x\sqrt{1-x^2}\, dx$

Let $u = 1 - x^2$

$$du = -2x\, dx$$

$$dx = -\frac{du}{2x}$$

When $x = 0$, $u = 1$.
When $x = 1$, $u = 0$.

$$I_1 = -\int_1^0 x\sqrt{u}\left(\frac{du}{2x}\right)$$

$$= \frac{1}{2}\int_0^1 \sqrt{u}\, du$$

$$= \frac{1}{2}\left[\frac{2}{3}u^{\frac{3}{2}}\right]_0^1$$

$$= \frac{1}{2}\left(\frac{2}{3}1^{\frac{3}{2}} - \frac{2}{3}0^{\frac{3}{2}}\right)$$

$$= \frac{1}{2}\left(\frac{2}{3}\right)$$

$$= \frac{1}{3}$$

b $I_n = \int_0^1 x^n \sqrt{1 - x^2}\, dx$

Using integration by parts:

$u = x^{n-1}, \; v' = x\sqrt{1 - x^2}$

$u' = (n-1)x^{n-2}$ and $v = -\dfrac{1}{2}\left(\dfrac{2}{3}\right)(1 - x^2)^{\frac{3}{2}} = -\dfrac{1}{3}(1 - x^2)^{\frac{3}{2}}$.

So $I_n = \left[x^{n-1}\left(-\dfrac{1}{3}(1 - x^2)^{\frac{3}{2}}\right)\right]_0^1 - \left(-\dfrac{n-1}{3}\right)\int_0^1 (1 - x^2)^{\frac{3}{2}} x^{n-2}\, dx$

$= \left[-\dfrac{1}{3} x^{n-1}(1 - x^2)^{\frac{3}{2}}\right]_0^1 + \dfrac{n-1}{3}\int_0^1 (1 - x^2)^{\frac{3}{2}} x^{n-2}\, dx$

$= \left[\left(-\dfrac{1}{3}[1][1 - 1]^{\frac{3}{2}} - \left(-\dfrac{1}{3}\right)[0][1 - 0]^{\frac{3}{2}}\right)\right] + \dfrac{n-1}{3}\int_0^1 (1 - x^2)^{\frac{1}{2}}(1 - x^2) x^{n-2}\, dx$

$= 0 + \dfrac{n-1}{3}\int_0^1 (1 - x^2)^{\frac{1}{2}}(x^{n-2} - x^n)\, dx$

$= \dfrac{n-1}{3}\int_0^1 \sqrt{1 - x^2}\,(x^{n-2} - x^n)\, dx$

$= \dfrac{n-1}{3}\int_0^1 x^{n-2}\sqrt{1 - x^2} - x^n\sqrt{1 - x^2}\, dx$

$= \dfrac{n-1}{3}\left(I_{n-2} - I_n\right)$

$3I_n = (n-1)I_{n-2} - (n-1)I_n$

$(3 + n - 1)I_n = (n-1)I_{n-2}$

$(n+2)I_n = (n-1)I_{n-2}$

$I_n = \left(\dfrac{n-1}{n+2}\right)I_{n-2}$, as required.

c $I_5 = \left(\dfrac{5-1}{5+2}\right)I_3$

$= \left(\dfrac{4}{7}\right)I_3$

$= \left(\dfrac{4}{7}\right)\left(\dfrac{3-1}{3+2}\right)I_1$

$= \left(\dfrac{4}{7}\right)\left(\dfrac{2}{5}\right)\left(\dfrac{1}{3}\right)$

$= \dfrac{8}{105}$

9780170459266

HSC exam topic grid (2011–2020)

This table shows the coverage of this topic in past HSC exams by question number. The past exams can be downloaded from the NESA website (www.educationstandards.nsw.edu.au) by selecting 'Year 11 – Year 12', 'HSC exam papers'. NESA marking feedback and guidelines can also be found there.

The new Mathematics Extension 2 course was first examined in 2020. For exams before 2020, select 'Year 11 – Year 12', 'Resources archive', 'HSC exam papers archive'.

	Integration by substitution	Rational functions and partial fractions	Integration by parts	Integration by recurrence relations
2011	1(b), 1(d), 7(b)	1(c), 1(e)	1(a), 8(a)	8(a)
2012	11(e), 12(a)	11(c), 14(a)	12(c)	12(c)
2013	**1**, 11(d), 12(a)	6, 11(b)	13(a)	13(a)
2014	**7**, 10, 13(a), 16(c)	1	11(b)	12(d)
2015	11(f)	11(c)	6, 14(a)	14(a)
2016	14(a)(i),(ii)	12(b), 15(c)(i)	11(b), 12(b)	14(b)
2017	11(d), 11(f)	14(a)	12(c), **15(a)**	**15(a)**
2018	14(a)	1, 11(c), 12(c)	**14(c)**	**14(c)**
2019	2, **15(a)**	11(c), 11(d), 15(c)	3, 15(c)	15(c)
2020 new course	**10**	6	11(b), 16(b)	16(b)

Questions in **bold** can be found in this chapter.

CHAPTER 5
MECHANICS

MECHANICS

Simple harmonic motion

- Velocity and acceleration as functions of displacement
- Equations and graphs for simple harmonic motion
- Acceleration, velocity and displacement
- Amplitude, period, phase shift, centre
- Properties of simple harmonic motion: maximum and zero velocity and acceleration

Modelling motion

- Newton's laws of motion: $F = m\ddot{x}$
- Forces and resolving forces
- Force diagrams
- Coefficient of friction

Resisted motion

- Resisted horizontal motion
- $R = kv$ or kv^2
- $v = f(t)$ and $v = f(x)$
- Resisted vertical motion under gravity and other forces
- Terminal velocity

Projectiles and resisted motion

- Equation of the path of a projectile
- Projectile motion under gravity and other forces

Glossary

acceleration
The rate of change of velocity with respect to time.

amplitude
The maximum distance travelled from the centre by a particle undergoing simple harmonic motion.

applied force
A force that is applied to an object by a person or another object; it can be a push or a pull action.

at rest
Stationary, at zero velocity.

displacement
A change in position relative to original position.

friction
The force exerted on an object by the surface it sits on or moves on.

initially
At the start, when time is zero.

newton
Symbol N. A metric unit of force, the amount needed to produce an acceleration of $1\,\mathrm{m\,s^{-2}}$ on a mass of $1\,\mathrm{kg}$.

normal force
For an object on a surface, the reactive force of the surface on that object that is equal in size to the force of the object on the surface and acting perpendicular to the surface.

period
The time it takes a particle to complete one cycle or oscillation when undergoing simple harmonic motion.

$$T = \frac{2\pi}{n}$$

phase shift
A horizontal translation of a trigonometric function.

projectile
An object that is thrown or projected upwards.

A+ DIGITAL FLASHCARDS
Revise this topic's key terms and concepts by scanning the QR code or typing the URL into your browser.
https://get.ga/a-hsc-maths-ext-2

range (or horizontal range)
The horizontal distance travelled by a projectile.

resisted motion
Motion that encounters resisting forces, for example, friction and air resistance.

resolving a force
To write a force in terms of its horizontal and vertical components $F\cos\theta$ and $F\sin\theta$, respectively.

simple harmonic motion
Motion in which an object's acceleration is proportional to and in the opposite direction to its displacement.

tension
The pulling force transmitted by means of a string, cable, chain or similar.

terminal velocity
The constant velocity that a free-falling object will eventually reach when the resistance of the medium through which the object is falling prevents further acceleration.

trajectory
The path of a projectile, which has the shape of a parabola when only gravity is acting on it.

velocity
The rate of change of displacement with respect to time.

weight
The amount of gravitational force acting on matter.

$$\text{weight} = \text{mass} \times \text{gravity}.$$

GLOSSARY

Topic summary

Applications of calculus to mechanics (MEX-M1)

M1.1 Simple harmonic motion

Velocity and acceleration as functions of x (displacement)

Velocity and acceleration are, by definition, functions of time:

$$v = \dot{x} = \frac{dx}{dt} \qquad a = \ddot{x} = \frac{dv}{dt}$$

However, if acceleration is determined by displacement, x, then we need to rewrite these functions as derivatives with respect to x.

$$a = \frac{d}{dx}\left(\frac{1}{2}v^2\right) \qquad a = v\frac{dv}{dx}$$

These formulas appear in the HSC exam reference sheet this way:

$$\frac{d^2x}{dt^2} = \frac{dv}{dt} = v\frac{dv}{dx} = \frac{d}{dx}\left(\frac{1}{2}v^2\right)$$

So if acceleration depends on time (t), write it as $a = \frac{dv}{dt}$, but if it depends on displacement (x), write it as $a = \frac{d}{dx}\left(\frac{1}{2}v^2\right)$ or $v\frac{dv}{dx}$.

Example 1

The acceleration of a particle is given by $\frac{d^2x}{dt^2} = 12x^2 - 2x + 3$, where x is the displacement.

Find the possible values of velocity when the particle is 5 m from the origin if initially the particle is at the origin and has velocity $-2\,\mathrm{m\,s}^{-1}$.

Solution

$$\frac{d^2x}{dt^2} = 12x^2 - 2x + 3$$

$$\frac{d}{dx}\left(\frac{1}{2}v^2\right) = 12x^2 - 2x + 3$$

$$\frac{1}{2}v^2 = 4x^3 - x^2 + 3x + c$$

When $t = 0$, $x = 0$, $v = -2$:

$$\frac{1}{2}(-2)^2 = 4(0)^3 - 0^2 + 3(0) + c$$

$$\therefore c = 2$$

So $\quad \dfrac{1}{2}v^2 = 4x^3 - x^2 + 3x + 2$

$$v^2 = 8x^3 - 2x^2 + 6x + 4.$$

When $x = 5$:

$$v^2 = 8(5^3) - 2(5^2) + 6(5) + 4$$

$$= 984$$

$$v = \pm\sqrt{984}$$

$$= \pm 31.4\,\mathrm{m\,s}^{-1}$$

Simple harmonic motion

Simple harmonic motion is a type of periodic motion or cyclic motion in which the restoring force is directly proportional to the displacement and acts in the direction opposite to that of the displacement. Simple harmonic motion is defined by

$$\ddot{x} = -n^2 x.$$

That is, the more you pull the particle one way, the more it 'wants' to return to the centre of the motion. The classic example is a mass on a spring because the more the mass stretches it, the more it wants to return towards the centre of the motion. Under simple harmonic motion, an object moves back and forth about a central position in a cyclic way. Examples include a mass attached to a light spring, a playground swing or pendulum (horizontally), the vibration of a guitar string (vertically), and the rise and fall of the tide in a river.

If a particle is undergoing simple harmonic motion about the origin, then

$$x = a \cos(nt + \alpha),$$

where a, n and α are constants and $a > 0$ and $n > 0$.

The **amplitude** (a) is the maximum value of x.

The **phase shift** $\left(\dfrac{\alpha}{n}\right)$ depends on the initial condition.

The **period** of the motion (T) is the time for the particle to complete one full oscillation (cycle),

$$T = \frac{2\pi}{n}.$$

> **Hint**
> Some places define phase shift as just α. Both are correct. It is more important to be able to use the value of α in a SHM problem. The syllabus uses $\dfrac{\alpha}{n}$ because
> $y = a \cos(nt + \alpha) = a \cos\left[n\left(t + \dfrac{\alpha}{n}\right)\right]$ is a
> horizontal translation of $y = a \cos t$ by $\dfrac{\alpha}{n}$ units.

The **frequency** (f) is the number of oscillations per second,

$$f = \frac{1}{T} = \frac{n}{2\pi}.$$

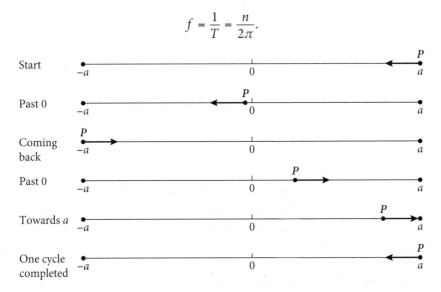

Proof

$$x = a \cos(nt + \alpha)$$
$$\dot{x} = -an \sin(nt + \alpha)$$
$$\ddot{x} = -an^2 \cos(nt + \alpha)$$
$$= -n^2 a \cos(nt + \alpha)$$
$$\text{Hence, } \ddot{x} = -n^2 x.$$

Note that the sine function $x = a \sin(nt + \alpha)$ can also be used to represent simple harmonic motion, and both the cosine and sine functions appear on the HSC exam reference sheet for simple harmonic motion (see the back of this book). It is better to use the sine function when the particle starts its motion at the centre, because $\sin 0 = 0$.

For $x = a \sin(nt + \alpha)$, we can also show that $\ddot{x} = -n^2 x$.

Example 2 ©NESA 2020 HSC EXAM, QUESTION 5

A particle undergoing simple harmonic motion has a maximum acceleration of $6\,m/s^2$ and a maximum velocity of $4\,m/s$.

What is the period of the motion?

A π **B** $\dfrac{2\pi}{3}$ **C** 3π **D** $\dfrac{4\pi}{3}$

Solution

$x = a\cos(nt + \alpha)$

$\dot{x} = -an\sin(nt + \alpha)$

$\ddot{x} = -an^2\cos(nt + \alpha)$

Max $\ddot{x} = 6$

$\therefore an^2 = 6$ [1]   when $\cos = -1$

Max $\ddot{x} = 4$

$\therefore an = 4$ [2]   when $\sin = 1$

$[1] \div [2]$:

$\dfrac{an^2}{an} = n = \dfrac{6}{4} = \dfrac{3}{2}$

$\text{Period} = \dfrac{2\pi}{n}$

$= \dfrac{2\pi}{\frac{3}{2}}$

$= \dfrac{4\pi}{3}$ (option D)

Velocity in simple harmonic motion

$$v^2 = n^2(a^2 - x^2)$$

Properties of simple harmonic motion

For simple harmonic motion oscillating about the origin,

$$\ddot{x} = -n^2 x \qquad v^2 = n^2(a^2 - x^2) \qquad x = a\cos(nt + \alpha)$$

	Equation	At $x = 0$ (centre)	At $x = \pm a$ (extremities)
Velocity	$\dot{x}^2 = n^2(a^2 - x^2)$	$\dot{x} = \pm na$ * Maximum velocity with the sign indicating the direction of motion.	$\dot{x} = 0$ Minimum velocity of 0, at rest.
Acceleration	$\ddot{x} = -n^2 x$	$\ddot{x} = 0$ Minimum acceleration (and force) of 0, changing signs.	$\ddot{x} = \mp n^2 a$ * Maximum acceleration with the sign indicating the direction of acceleration (and force), always towards centre.

*The formulas for maximum velocity and acceleration can also be determined from the sine and cosine formulas for $\dot{x}$ and $\ddot{x}$, as seen in Example 2.

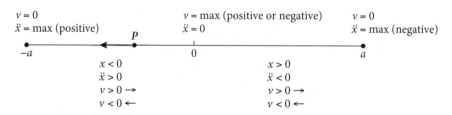

- At $x = 0$, $v = \pm na$. The velocity of the particle has maximum magnitude at the centre of motion with the sign indicating the direction of motion.

- At $x = \pm a$, $v = 0$. The velocity of the particle is zero at the extremities of the motion.

- At $x = 0$, $\ddot{x} = 0$. The acceleration of the particle is zero at the centre of the motion.

- At $x = \pm a$, $\ddot{x} = \mp n^2 a$. The magnitude of acceleration is a maximum at the extremities of the motion and is always directed towards the centre of the motion.

Simple harmonic motion about $x = c$

If the particle is undergoing simple harmonic motion about $x = c$ rather than $x = 0$, then the equations of motion are translated where x is replaced by $x - c$.

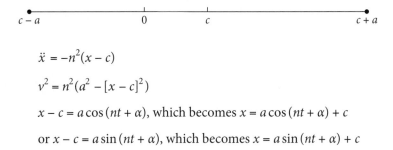

$$\ddot{x} = -n^2(x - c)$$

$$v^2 = n^2(a^2 - [x - c]^2)$$

$x - c = a\cos(nt + \alpha)$, which becomes $x = a\cos(nt + \alpha) + c$

or $x - c = a\sin(nt + \alpha)$, which becomes $x = a\sin(nt + \alpha) + c$

Example 3 ©NESA 2020 HSC EXAM, QUESTION 13(a)

A particle is undergoing simple harmonic motion with period $\dfrac{\pi}{3}$. The central point of motion of the particle is at $x = \sqrt{3}$. When $t = 0$ the particle has its maximum displacement of $2\sqrt{3}$ from the central point of motion.

Find an equation for the displacement, x, of the particle in terms of t.

Solution

Let $x = a\cos(nt + \alpha) + c$.

Period $= \dfrac{2\pi}{n} = \dfrac{\pi}{3}$

$$6\pi = n\pi$$

$$n = \dfrac{6\pi}{\pi} = 6$$

$a = $ amplitude $= 2\sqrt{3}$

$c = $ central point $= \sqrt{3}$

$x = 2\sqrt{3}\cos(6t + \alpha) + \sqrt{3}$

When $t = 0$, $x = 3\sqrt{3}$ at endpoint:

$$3\sqrt{3} = 2\sqrt{3}\cos\alpha + \sqrt{3}$$
$$2\sqrt{3} = 2\sqrt{3}\cos\alpha$$
$$1 = \cos\alpha$$
$$\alpha = 0$$

So $x = 2\sqrt{3}\cos 6t + \sqrt{3}$.

Example 4 ©NESA 2016 MATHEMATICS EXTENSION 1 HSC EXAM, QUESTION 13(a)

The tide can be modelled using simple harmonic motion. At a particular location, the high tide is 9 metres and the low tide is 1 metre. At this location, the tide completed 2 full periods every 25 hours.

Let t be the time in hours after the first high tide today.

a Explain why the tide can be modelled by the function $x = 5 + 4\cos\left(\frac{4\pi}{25}t\right)$.

b The first high tide tomorrow is at 2 am.

What is the earliest time tomorrow at which the tide is increasing at the fastest rate?

Solution

a Let $x = a\cos(nt + \alpha) + c$.

High tide = 9 m, low tide = 1 m

Centre, $c = \frac{1+9}{2} = 5$

Amplitude, $a = 9 - 5 = 4$ $\left(\text{or } \frac{9-1}{2}\right)$

Period $= \frac{2\pi}{n} = \frac{25}{2}$

$4\pi = 25n$

$n = \frac{4\pi}{25}$

Substituting these values in $x = a\cos(nt + \alpha) + c$:

$$x = 4\cos\left(\frac{4\pi}{25}t + \alpha\right) + 5$$

When $t = 0$, $x = 9$ at high tide: (or $\cos 0$)

$$9 = 4\cos(0 + \alpha) + 5$$
$$4 = 4\cos\alpha$$
$$1 = \cos\alpha$$
$$\alpha = 0$$

$\therefore x = 4\cos\left(\frac{4\pi}{25}t\right) + 5$, as required.

b 'Increasing at the fastest rate' means $v > 0$ and v is at a maximum, so $\ddot{x} = \frac{dv}{dt} = 0$.

Thinking about simple harmonic motion, this occurs at the centre of motion, when x is increasing, so at point T on the graph below, which shows 1 cycle of the tide.

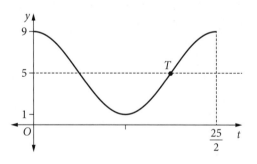

Or algebraically:

$$\ddot{x} = -n^2(x - c) = 0$$
$$x = c = 5$$

Substitute $x = 5$ into $x = 4\cos\left(\frac{4\pi}{25}t\right) + 5$:

$$5 = 4\cos\left(\frac{4\pi}{25}t\right) + 5$$
$$0 = 4\cos\left(\frac{4\pi}{25}t\right)$$
$$\cos\left(\frac{4\pi}{25}t\right) = 0$$
$$\frac{4\pi}{25}t = \frac{\pi}{2}, \frac{3\pi}{2}, \frac{5\pi}{2}, \ldots$$
$$t = \frac{25}{8}, \frac{75}{8}, \frac{125}{8}, \ldots$$

Thinking about the cosine graph, the tide is *decreasing* fastest at $\frac{25}{8}$ hours after high tide and *increasing* fastest at $\frac{75}{8}$ hours after high tide, at point T.

Time = 2 am + $\frac{75}{8}$ hours
    = 2 am + 9 hours 22.5 min
    ≈ 11:22 am

$\therefore$ the earliest time tomorrow at which the tide is increasing at the fastest rate is 11:22 am.

M1.2 Modelling motion without resistance

Newton's 3 laws of motion

1. An object remains in a state of rest or uniform motion in a straight line unless it is acted upon by an external force.

2. For a constant mass, force = mass × acceleration.

3. For every action, there is an equal and opposite reaction.

Newton's first law defines inertia, and states that if there is no net force acting on an object then the object will remain at constant velocity (including zero velocity if it was not already moving and the object remains at rest).

Force

Force is measured in newtons (N). One newton is the force required to produce an acceleration of $1 \, \text{m s}^{-2}$ on a mass of 1 kg.

Normal force is the support force exerted upon an object that is in contact with another stable object. For example, when a book is on a desk, they exert a normal force on each other, perpendicular to the contacting surfaces.

Friction is the resistance to motion of one object moving relative to another. Friction opposes the direction of motion. If 2 surfaces can move over each other without any resistance, they have smooth contact.

Friction is proportional to the normal reaction on the object, related by the coefficient of friction, μ (mu).

$$F = \mu N$$

The coefficient of friction depends on the material of the objects in contact. The **static coefficient of friction** is applied when both objects are not moving. The **dynamic coefficient of friction** is applied when one or both objects are moving.

Weight is the force due to gravity.

Weight = mass × gravity.

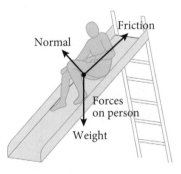

Tension on an object is transmitted through a string, rope or wire when it is pulled tight from opposite ends. The tension is directed along the length of the wire and pulls equally on the objects on the opposite ends of the wire.

Applied force is exerted on an object by a person or another object.

When all forces are equal and opposite, the object remains at rest. If, however, the forces are not in balance, then the object will accelerate in the direction determined by the net resulting force.

Resolving forces

If a force (F) makes an angle of θ with the x-axis, then F can be written in terms of components in the x and y directions.

$F\cos\theta$ is the horizontal component and $F\sin\theta$ is the vertical component. The process of writing a force in terms of its horizontal and vertical components is called **resolving a force**.

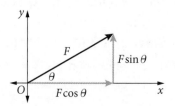

Example 5

A 40 N force is applied to the 5 kg box at an angle of 30° to the horizontal. If the dynamic coefficient of friction is 0.4, find the acceleration of the box.

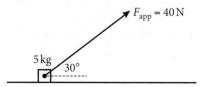

Solution

A complete force diagram:

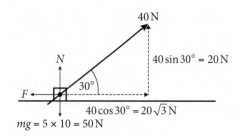

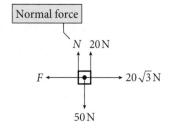

Resolving forces vertically:

Normal force $(N) = 50 - 20$
$$= 30 \text{ newtons}$$

Resolving forces horizontally:

$F = \mu N$,   where μ is the dynamic coefficient of friction.
$$= 0.4 \times 30$$
$$= 12 \text{ newtons}$$

> **Hint**
> N on its own means 'normal force', but N after a number means 'newtons', such as '40 N'.

Net force horizontally:

$$F_{net} = 20\sqrt{3} - 12$$
$$= 22.641\ldots \text{ newtons}$$

For the acceleration of the box:

$$F_{net} = m\ddot{x}$$
$$22.641\ldots = 5\ddot{x}$$
$$\ddot{x} \approx 4.5 \text{ m s}^{-2}$$

Example 6

Calculate, correct to the nearest newton, the tension in a wire supporting a 60 kg tightrope walker whose weight at the centre of the wire causes it to sag by 8° to the horizontal. Let $g = 9.8 \text{ m s}^{-2}$.

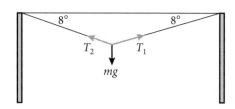

Solution

Horizontally:

$$T_1 \cos 8° = T_2 \cos 8°$$

Hence, $T_1 = T_2$.

Vertically:

$$T_1 \sin 8° + T_2 \sin 8° = mg$$

Let $T = T_1 = T_2$:

$$T\sin 8° + T\sin 8° = mg$$
$$2T\sin 8° = mg$$
$$T = \frac{mg}{2\sin 8°}$$
$$= \frac{60 \times 9.8}{2\sin 8°}$$
$$= 2112.4771\ldots$$
$$\approx 2112 \text{ N}$$

Example 7   ©NESA   2020 HSC EXAM, QUESTION 12(a)

A 50-kilogram box is initially at rest. The box is pulled along the ground with a force of 200 newtons at an angle of $30°$ to the horizontal. The box experiences a resistive force of $0.3R$ newtons, where R is the normal force, as shown in the diagram.

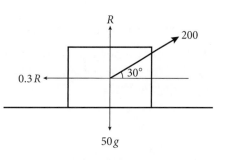

Take the acceleration g due to gravity to be $10\,\text{m/s}^2$.

a By resolving the forces vertically, show that $R = 400$.

b Show that the net force horizontally is approximately 53.2 newtons.

c Find the velocity of the box after the first three seconds.

> **Hint**
> g means 'gravitational acceleration' here, not 'grams'.

Solution

a Resolving forces vertically:

$$R + 200 \sin 30° = 50g$$
$$R + 200 \times \frac{1}{2} = 50 \times 10$$
$$R + 100 = 500$$
$$R = 400$$

b Net horizontal force $(m\ddot{x}) = 200 \cos 30° - 0.3R$

$$= \frac{200\sqrt{3}}{2} - 0.3 \times 400$$
$$= 100\sqrt{3} - 120$$
$$= 53.2050\ldots$$
$$\approx 53.2\,\text{N}$$

c $F = m\ddot{x}$

$$50\ddot{x} = 53.2$$
$$\ddot{x} = \frac{53.2}{50}$$
$$= 1.064$$

$$\frac{dv}{dt} = 1.064$$
$$v = 1.064t + c$$

When $t = 0$, $v = 0$:

$$0 = 0 + c$$
$$\therefore c = 0$$

$$\therefore v = 1.064t$$

Velocity after 3 seconds:

$$v = 1.064 \times 3$$
$$= 3.192\,\text{m s}^{-1}$$

M1.3 Resisted motion

Resisted horizontal motion

When an object moves through air, water or oil, the resistive force (R) of the medium slows down the object and is proportional to the object's velocity or velocity squared:

$$R = -kv \text{ or } R = -kv^2$$

$\longrightarrow$ Direction of motion

$R \longleftarrow \bullet$

Example 8

A particle of mass m initially with speed v_0 moves horizontally against a resistance proportional to the square of the speed.

Find the equation of its speed in terms of its displacement.

Solution

$m\ddot{x} = -kv^2$

$\ddot{x} = -\dfrac{k}{m}v^2$

$v\dfrac{dv}{dx} = -\dfrac{k}{m}v^2$

$\dfrac{dv}{dx} = -\dfrac{k}{m}v$

$\dfrac{dx}{dv} = -\dfrac{m}{kv}$

$x = -\dfrac{m}{k}\ln v + c, \ v > 0$ because it is speed.

When $x = 0, \ v = v_0$:

$0 = -\dfrac{m}{k}\ln v_0 + c$

$c = \dfrac{m}{k}\ln v_0$

$x = -\dfrac{m}{k}\ln v + \dfrac{m}{k}\ln v_0$

$x = \dfrac{m}{k}(\ln v_0 - \ln v)$

$\dfrac{kx}{m} = \ln\left(\dfrac{v_0}{v}\right)$

$e^{\frac{kx}{m}} = \dfrac{v_0}{v}$

$v = \dfrac{v_0}{e^{\frac{kx}{m}}}$

$v = v_0 e^{-\frac{kx}{m}}$

Example 9 ©NESA 2020 HSC EXAM, QUESTION 14(b)

A particle starts from rest and falls through a resisting medium so that its acceleration, in m/s^2, is modelled by

$$a = 10(1 - (kv)^2),$$

where v is the velocity of the particle in m/s and $k = 0.01$.

Find the velocity of the particle after 5 seconds.

Solution

$a = 10(1 - [kv]^2)$

$\dfrac{dv}{dt} = 10(1 - k^2v^2)$

$\dfrac{dt}{dv} = \dfrac{1}{10(1 - k^2v^2)}$

Separate $\dfrac{1}{1 - k^2v^2}$ into partial fractions:

$\dfrac{1}{1 - k^2v^2} = \dfrac{1}{(1 - kv)(1 + kv)} = \dfrac{A}{1 - kv} + \dfrac{B}{1 + kv}$

$A(1 + kv) + B(1 - kv) = 1$

$\quad A + Akv + B - Bkv = 1$

$\quad A + B + kv(A - B) = 1$

$A + B = 1 \qquad A - B = 0$

Add them:

$2A = 1$

$A = \dfrac{1}{2}$

$B = 1 - \dfrac{1}{2} = \dfrac{1}{2}$

$\therefore \dfrac{1}{1 - k^2v^2} = \dfrac{1}{2}\left(\dfrac{1}{1 - kv} + \dfrac{1}{1 + kv}\right)$

$\dfrac{dt}{dv} = \dfrac{1}{10(1 - k^2v^2)}$

$\quad = \dfrac{1}{10}\dfrac{1}{2}\left(\dfrac{1}{1 - kv} + \dfrac{1}{1 + kv}\right)$

$\quad = \dfrac{1}{20}\left(\dfrac{1}{1 - kv} + \dfrac{1}{1 + kv}\right)$

$t = \dfrac{1}{20}\displaystyle\int \dfrac{1}{1 - kv} + \dfrac{1}{1 + kv}\, dv$

$20t = -\dfrac{1}{k}\ln|1 - kv| + \dfrac{1}{k}\ln|1 + kv| + c$

When $t = 0$, $v = 0$:

$0 = -\dfrac{1}{k}\ln 1 + \dfrac{1}{k}\ln 1 + c$

$0 = 0 + 0 + c$

$20t = -\dfrac{1}{k}\ln|1 - kv| + \dfrac{1}{k}\ln|1 + kv|$

$20kt = -\ln|1 - kv| + \ln|1 + kv|$

$\ln\left|\dfrac{1 + kv}{1 - kv}\right| = 20kt$

$\dfrac{1 + kv}{1 - kv} = \pm e^{20kt}$

By considering the starting conditions $t = 0$, $v = 0$, we can see that LHS = 1, RHS = ±1, so the positive solution is correct here.

(Alternatively, $a > 0$, so $1 - [kv]^2 > 0$, so $[kv]^2 < 1$, so $0 < kv < 1$ as k, $v > 0$, so $\dfrac{1 + kv}{1 - kv} > 0$).

Substitute $k = 0.01$, $t = 5$ to find v:

$\dfrac{1 + 0.01v}{1 - 0.01v} = e^{20(0.01)5}$

$\dfrac{1 + 0.01v}{1 - 0.01v} = e^1$

$1 + 0.01v = e(1 - 0.01v)$

$\qquad\qquad = e - 0.01ev$

$0.01v + 0.01ev = e - 1$

$\quad 0.01v(1 + e) = e - 1$

$\qquad\qquad v = \dfrac{e - 1}{0.01(1 + e)}$

$\qquad\qquad\quad \approx 46.2\,\text{m/s}$

Resisted vertical motion

When a particle is moving vertically (either upwards or downwards), the acceleration due to gravity is always towards Earth. The acceleration is represented by g, and near Earth's surface, $g \approx 9.8\,\mathrm{m\,s^{-2}}$, often rounded to $10\,\mathrm{m\,s^{-2}}$.

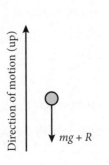
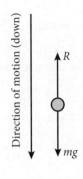

There may also be a resistance, R, to the particle whose direction is always opposing the direction of motion. Again,

$$R = -kv \text{ or } R = -kv^2.$$

Terminal velocity

As an object falls under gravity and increases in velocity, the air resistance acting against it also increases as it is dependent on the velocity. At some point, the air resistance will be equal to the force of gravity. When this happens, there is zero net force acting and the object is no longer accelerating. It cannot fall any faster, and its velocity has reached **terminal velocity,** v_T, its maximum speed.

$$v_T \text{ occurs when acceleration } \ddot{x} = 0 \text{ as } t \to \infty.$$

Example 10   ©NESA   2019 HSC EXAM, QUESTION 14(b)

A parachutist jumps from a plane, falls freely for a short time and then opens the parachute. Let t be the time in seconds after the parachute opens, $x(t)$ be the distance in metres travelled after the parachute opens, and $v(t)$ be the velocity of the parachutist in $\mathrm{m\,s^{-1}}$.

The acceleration of the parachutist after the parachute opens is given by

$$\ddot{x} = g - kv,$$

where $g\,\mathrm{m\,s^{-2}}$ is the acceleration due to gravity and k is a positive constant.

a   With an open parachute the parachutist has a terminal velocity of $w\,\mathrm{m\,s^{-1}}$.

Show that $w = \dfrac{g}{k}$.

At the time the parachute opens, the speed of descent is $1.6w\,\mathrm{m\,s^{-1}}$.

b   Show that it takes $\dfrac{1}{k}\log_e 6$ seconds to slow down to a speed of $1.1w\,\mathrm{m\,s^{-1}}$.

c   Let D be the distance the parachutist travels between opening the parachute and reaching the speed $1.1w\,\mathrm{m\,s^{-1}}$.

Show that $D = \dfrac{g}{k^2}\left(\dfrac{1}{2} + \log_e 6\right)$.

Solution

a   For terminal velocity, $\ddot{x} = 0$.

$$\ddot{x} = g - kv = 0$$
$$g = kv$$
$$v = \frac{g}{k}$$

So $w = \dfrac{g}{k}$.

b Need to find v in terms of t.

$$\ddot{x} = \frac{dv}{dt} = g - kv$$

$$\frac{dt}{dv} = \frac{1}{g - kv}$$

$$t = -\frac{1}{k}\ln|g - kv| + c$$

When parachute opens,

$\ddot{x} = g - kv < 0$ (deceleration),
so $|g - kv| = kv - g$.

OR

As $v > \frac{g}{k}$ (terminal velocity), $kv > g$,

so $g - kv < 0$.

$$t = -\frac{1}{k}\ln(kv - g) + c$$

When $t = 0$, $v = 1.6w = \frac{1.6g}{k}$:

$$0 = -\frac{1}{k}\ln\left[k\left(\frac{1.6g}{k}\right) - g\right] + c$$

$$= -\frac{1}{k}\ln(1.6g - g) + c$$

$$= -\frac{1}{k}\ln(0.6g) + c$$

$$c = \frac{1}{k}\ln(0.6g)$$

$$t = -\frac{1}{k}\ln(kv - g) + \frac{1}{k}\ln(0.6g)$$

$$= \frac{1}{k}\ln\left(\frac{0.6g}{kv - g}\right)$$

When $v = 1.1w = \frac{1.1g}{k}$:

$$t = \frac{1}{k}\ln\left(\frac{0.6g}{k\left[\frac{1.1g}{k}\right] - g}\right)$$

$$= \frac{1}{k}\ln\left(\frac{0.6g}{1.1g - g}\right)$$

$$= \frac{1}{k}\ln\left(\frac{0.6g}{0.1g}\right)$$

$$= \frac{1}{k}\ln 6, \text{ as required.}$$

c Need to find x in terms of t, and then

substitute $t = \frac{1}{k}\log_e 6$.

$$t = \frac{1}{k}\ln\left(\frac{0.6g}{kv - g}\right)$$

Make v the subject, then integrate:

$$kt = \ln\left(\frac{0.6g}{kv - g}\right)$$

$$e^{kt} = \frac{0.6g}{kv - g}$$

$$kv - g = \frac{0.6g}{e^{kt}} = 0.6ge^{-kt}$$

$$kv = 0.6ge^{-kt} + g$$

$$v = \frac{g}{k}(0.6e^{-kt} + 1)$$

$$x = \frac{g}{k}\left(-\frac{0.6}{k}e^{-kt} + t\right) + d$$

When $t = 0$, $x = 0$:

$$0 = \frac{g}{k}\left(-\frac{0.6}{k}[1] + 0\right) + d$$

$$0 = -\frac{0.6g}{k^2} + d$$

$$d = \frac{0.6g}{k^2}$$

$$x = \frac{g}{k}\left(t - \frac{0.6}{k}e^{-kt}\right) + \frac{0.6g}{k^2}$$

$$= \frac{g}{k}\left(t - \frac{0.6}{k}e^{-kt} + \frac{0.6}{k}\right)$$

When $t = \frac{1}{k}\ln 6$, $x = D$:

$$D = \frac{g}{k}\left(\frac{1}{k}\ln 6 - \frac{0.6}{k}e^{-k\left[\frac{1}{k}\ln 6\right]} + \frac{0.6}{k}\right)$$

$$= \frac{g}{k^2}(\ln 6 - 0.6e^{-\ln 6} + 0.6)$$

$$= \frac{g}{k^2}(\ln 6 - 0.6[e^{\ln 6}]^{-1} + 0.6)$$

$$= \frac{g}{k^2}(\ln 6 - 0.6[6]^{-1} + 0.6)$$

$$= \frac{g}{k^2}(\ln 6 - 0.1 + 0.6)$$

$$= \frac{g}{k^2}(\ln 6 + 0.5)$$

$$= \frac{g}{k^2}\left(\frac{1}{2} + \ln 6\right), \text{ as required.}$$

Example 11 ©NESA 2018 HSC EXAM, QUESTION 14(b)

A falling particle experiences forces due to gravity and air resistance. The acceleration of the particle is $g - kv^2$, where g and k are positive constants and v is the speed of the particle. (Do NOT prove this.)

Prove that, after falling from rest through a distance, h, the speed of the particle will be

$$\sqrt{\frac{g}{k}(1 - e^{-2kh})}.$$

Solution

$\ddot{x} = g - kv^2$

Need to find an equation for v in terms of x.

$v\dfrac{dv}{dx} = g - kv^2$

$\dfrac{dv}{dx} = \dfrac{g - kv^2}{v}$

$\dfrac{dx}{dv} = \dfrac{v}{g - kv^2} = -\dfrac{1}{2k}\dfrac{-2kv}{g - kv^2}$

$x = -\dfrac{1}{2k}\ln\left|g - kv^2\right| + c$

But $\ddot{x} = g - kv^2 > 0$,

so $x = -\dfrac{1}{2k}\ln(g - kv^2) + c$.

When $x = 0$, $v = 0$:

$0 = -\dfrac{1}{2k}\ln(g - 0) + c$

$c = \dfrac{1}{2k}\ln g$

$x = -\dfrac{1}{2k}\ln(g - kv^2) + \dfrac{1}{2k}\ln g$

$\quad = \dfrac{1}{2k}\ln\left(\dfrac{g}{g - kv^2}\right)$

When $x = h$,

$h = \dfrac{1}{2k}\ln\left(\dfrac{g}{g - kv^2}\right)$

$2kh = \ln\left(\dfrac{g}{g - kv^2}\right)$

$e^{2kh} = \dfrac{g}{g - kv^2}$

$g - kv^2 = \dfrac{g}{e^{2kh}} = ge^{-2kh}$

$g - ge^{-2kh} = kv^2$

$kv^2 = g(1 - e^{-2kh})$

$v^2 = \dfrac{g}{k}(1 - e^{-2kh})$

Speed $v = \sqrt{\dfrac{g}{k}(1 - e^{-2kh})}$, as required.

M1.4 Projectiles and resisted motion

Projectile motion

Projectile motion, the study of free fall under gravity in 2 dimensions rather than just vertically, is mostly covered in the Mathematics Extension 1 course.

The path followed by a projectile is known as its **trajectory.** If the motion is not resisted, gravity forces the projectile to travel the path of a **parabola**, and gravity accelerates the object downwards.

The factors that affect the trajectory are:

- the angle of projection, θ

- the initial speed, V

- the starting point

This initial velocity, V, has 2 components, $\dot{x} = V\cos\theta$ and $\dot{y} = V\sin\theta$, which are velocity components in the x and y directions (or horizontal and vertical directions), respectively.

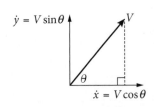

The horizontal component remains constant as there are no forces acting on the projectile in this direction and so it is independent of t, while the vertical component changes with t due to gravity.

Equations of projectile motion (starting point (0, 0))

	Horizontal motion	**Vertical motion**
Acceleration	$\ddot{x} = 0$	$\ddot{y} = -g$
Velocity	Integrating:	Integrating:
	$\dot{x} = c$	$\dot{y} = -gt + k$
	When $t = 0$, $\dot{x} = V\cos\theta$:	When $t = 0$, $\dot{y} = V\sin\theta$:
	$V\cos\theta = c$	$V\sin\theta = k$
	$\therefore \dot{x} = V\cos\theta$	$\therefore \dot{y} = -gt + V\sin\theta$
Displacement	Integrating:	Integrating:
	$x = Vt\cos\theta + d$	$y = -\dfrac{1}{2}gt^2 + Vt\sin\theta + l$
	When $t = 0$, $x = 0$:	
	$0 = 0 + d$	When $t = 0$, $y = 0$:
	$d = 0$	$0 = 0 + 0 + l$
	$\therefore x = Vt\cos\theta$	$l = 0$
		$y = -\dfrac{1}{2}gt^2 + Vt\sin\theta$

The equations for velocity and displacement do not need to be memorised, but they must be proved for each problem. The same applies to the following formulas.

Maximum height

For maximum height $\dot{y} = 0$.

Hence, $0 = -gt + V\sin\theta$

$$gt = V\sin\theta$$

$$t = \frac{V\sin\theta}{g} \text{ is the time to reach maximum height.}$$

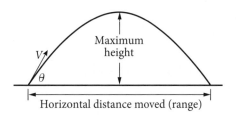

Substitute into y, vertical displacement:

$$y = -\frac{1}{2}g\left(\frac{V\sin\theta}{g}\right)^2 + V\left(\frac{V\sin\theta}{g}\right)\sin\theta$$

$$y = -\frac{V^2\sin^2\theta}{2g} + \frac{V^2\sin^2\theta}{g}$$

$$y = -\frac{V^2\sin^2\theta}{2g} + \frac{2V^2\sin^2\theta}{2g}$$

$$\text{So } y_{\text{max}} = \frac{V^2\sin^2\theta}{2g}.$$

Time of flight

Projectile returns to the ground when $y = 0$.

$$0 = -\frac{1}{2}gt^2 + Vt\sin\theta$$

$$0 = -gt^2 + 2Vt\sin\theta$$

$$gt^2 = 2Vt\sin\theta$$

$$gt = 2V\sin\theta \quad (t \neq 0)$$

$$t = \frac{2V\sin\theta}{g}$$

Note that the time of flight is double the time to reach the maximum height $t = \dfrac{V\sin\theta}{g}$.

Range (horizontal distance travelled)

Substitute $t = \dfrac{2V \sin \theta}{g}$ into x:

$$x = V\left(\frac{2V \sin \theta}{g}\right)\cos \theta$$

$$x = \frac{2V^2 \sin \theta \cos \theta}{g}$$

$$x = \frac{V^2 \sin 2\theta}{g}$$

Maximum range

For maximum range, we want $\sin 2\theta = 1$.

$$x_{\text{max}} = \frac{V^2}{g}$$

This occurs when the angle is $45°$.

$$\sin 2\theta = 1$$
$$2\theta = 90°$$
$$\theta = 45°$$

Speed at any point

The speed at any point on the trajectory depends on both the horizontal and vertical components of velocity.

$$S = \sqrt{\dot{x}^2 + \dot{y}^2}$$

Equation of the trajectory (path of the projectile)

$x = Vt \cos \theta$ and $y = -\dfrac{1}{2}gt^2 + Vt \sin \theta$ are parametric equations of t. We can eliminate t to get a Cartesian equation for y in terms of x, the equation of the trajectory.

From the equation for x:

$$t = \frac{x}{V \cos \theta}$$

Substitute into y:

$$y = -\frac{1}{2}g\left(\frac{x}{V \cos \theta}\right)^2 + V\left(\frac{x}{V \cos \theta}\right)\sin \theta$$

$$= -\frac{gx^2}{2V^2 \cos^2 \theta} + x\left(\frac{\sin \theta}{\cos \theta}\right)$$

$$y = \left(-\frac{g}{2V^2 \cos^2 \theta}\right)x^2 + (\tan \theta)x$$

This is the equation of a parabola of the form $y = ax^2 + bx$, concave down ($a < 0$).

Projection from a height, h, above the origin

In the case of a projectile launched from a height, h, the horizontal and vertical equations of motion do not change, except for the vertical displacement.

$$y = -\frac{1}{2}gt^2 + Vt \sin \theta + h$$

This is just the standard vertical displacement function translated h units upwards.

Example 12

A cannon is fired from a platform of height 25 metres. The cannonball has initial velocity 60 m/s at an angle of 48° to the horizontal. Assume $g = 10 \text{ m/s}^2$.

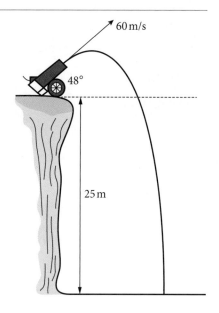

a How long (correct to two decimal places) will the cannonball take to land?

b Find the range of the cannon, correct to the nearest metre.

c What is the maximum height obtained by the cannonball, correct to the nearest metre?

TOPIC SUMMARY

Solution

a Vertical motion:

$$\ddot{y} = -g = -10$$
$$\dot{y} = -10t + k$$

When $t = 0$:

$$\dot{y} = 60 \sin 48°$$
$$60 \sin 48° = k$$
$$\therefore \dot{y} = -10t + 60 \sin 48°$$
$$y = -5t^2 + 60t \sin 48° + l$$

When $t = 0$, $y = 25$:

$$25 = 0 + 0 + l$$
$$l = 25$$
$$y = -5t^2 + 60t \sin 48° + 25$$

Time of flight is t when $y = 0$:

$$0 = -5t^2 + 60t \sin 48° + 25$$

Solve using the quadratic formula:

$$t = \frac{-60 \sin 48° \pm \sqrt{(60 \sin 48°)^2 - 4(-5)(25)}}{2(-5)}$$

$$\approx -0.53 \text{ or } 9.45$$
$$\approx 9.45 \text{ s} \quad \text{since } t > 0.$$

b Horizontal motion:

$$\ddot{x} = 0$$
$$\dot{x} = c$$

When $t = 0$:

$$\dot{x} = 60 \cos 48°:$$
$$60 \cos 48° = c$$
$$\therefore \dot{x} = 60 \cos 48°$$
$$x = 60t \cos 48° + d$$

When $t = 0$, $x = 0$:

$$0 = 0 + d$$
$$d = 0$$
$$\therefore x = 60t \cos 48°$$

For range, substitute $t = 9.45$:

$$x = Vt \cos \theta$$
$$= 60(9.45) \cos 48°$$
$$\approx 379 \text{ m}$$

c Maximum height is y when $\dot{y} = 0$:

$$0 = -10t + 60 \sin 48°$$
$$10t = 60 \sin 48°$$
$$t = \frac{60 \sin 48°}{10}$$
$$\approx 4.46 \text{ s}$$

Substitute into y:

$$y = -5(4.46)^2 + 60(4.46) \sin 48° + 25$$
$$\approx 124 \text{ m}$$

Example 13 ©NESA 2020 HSC EXAM, QUESTION 12(b)

A particle is projected from the origin with initial velocity u m/s at an angle θ to the horizontal. The particle lands at $x = R$ on the x-axis. The acceleration vector is given by $\underset{\sim}{a} = \begin{pmatrix} 0 \\ -g \end{pmatrix}$, where g is the acceleration due to gravity. (Do NOT prove this).

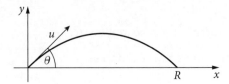

a Show that the position vector $\underset{\sim}{r}(t)$ of the particle is given by

$$\underset{\sim}{r}(t) = \begin{pmatrix} ut \cos \theta \\ ut \sin \theta - \dfrac{1}{2} gt^2 \end{pmatrix}.$$

b Show that the Cartesian equation of the path of flight is given by

$$y = \frac{-gx^2}{2u^2} \left(\tan^2 \theta - \frac{2u^2}{gx} \tan \theta + 1 \right).$$

c Given $u^2 > gR$, prove that there are 2 distinct values of θ for which the particle will land at $x = R$.

Solution

a $\underset{\sim}{a}(t) = \begin{pmatrix} 0 \\ -g \end{pmatrix}$

> **Hint**
> Here we are deriving the equations of motion in *vector form*, working on horizontal and vertical components simultaneously.

$\underset{\sim}{v}(t) = \begin{pmatrix} c_1 \\ -gt + c_2 \end{pmatrix}$

When $t = 0$, $\underset{\sim}{v}(0) = \begin{pmatrix} u \cos \theta \\ u \sin \theta \end{pmatrix}.$

So $u \cos \theta = c_1$
   $u \sin \theta = 0 + c_2.$

$\therefore \underset{\sim}{v}(t) = \begin{pmatrix} 0 + u \cos \theta \\ -gt + u \sin \theta \end{pmatrix}$

$\underset{\sim}{r}(t) = \begin{pmatrix} ut \cos \theta + c_3 \\ -\dfrac{1}{2} gt^2 + ut \sin \theta + c_4 \end{pmatrix}$

When $t = 0$, $\underset{\sim}{r}(0) = \begin{pmatrix} 0 \\ 0 \end{pmatrix}.$

So $0 = 0 + c_3$
   $0 = 0 + 0 + c_4.$

Hence, $\underset{\sim}{r}(t) = \begin{pmatrix} ut \cos \theta \\ ut \sin \theta - \dfrac{1}{2} gt^2 \end{pmatrix}$, as required.

b $x = ut \cos \theta \rightarrow t = \dfrac{x}{u \cos \theta}$

Substitute t into y-equation:

$y = ut \sin \theta - \dfrac{1}{2} gt^2$

$y = u \left(\dfrac{x}{u \cos \theta} \right) \sin \theta - \dfrac{1}{2} g \left(\dfrac{x}{u \cos \theta} \right)^2$

$= x \tan \theta - \dfrac{gx^2}{2u^2} (\sec^2 \theta)$

$= x \tan \theta - \dfrac{gx^2}{2u^2} (1 + \tan^2 \theta)$

$= -\dfrac{gx^2}{2u^2} \left(-\dfrac{2u^2}{gx} \tan \theta + 1 + \tan^2 \theta \right)$

$= -\dfrac{gx^2}{2u^2} \left(\tan^2 \theta - \dfrac{2u^2}{gx} \tan \theta + 1 \right)$

c When $x = R$, this quadratic equation in $\tan\theta$ has 2 distinct values.

$$0 = -\frac{gR^2}{2u^2}\left(\tan^2\theta - \frac{2u^2}{gR}\tan\theta + 1\right)$$

$\therefore \tan^2\theta - \frac{2u^2}{gx}\tan\theta + 1 = 0$ and $\Delta > 0$.

$$\Delta = b^2 - 4ac$$

$$= \left(-\frac{2u^2}{gR}\right)^2 - 4 \times 1 \times 1$$

$$= \frac{4u^4}{g^2R^2} - 4$$

$$= \frac{4u^4 - 4g^2R^2}{g^2R^2}$$

$$= \frac{4(u^4 - g^2R^2)}{g^2R^2}$$

As g and R are positive:

$$u^2 > gR > 0$$
$$\therefore u^4 > g^2R^2 > 0$$
$$u^4 - g^2R^2 > 0$$

$$\therefore \Delta = \frac{4(u^4 - g^2R^2)}{g^2R^2} > 0$$

So the quadratic equation has 2 distinct values and there are 2 distinct values for θ for which the particle will land at $x = R$.

Resisted projectile motion

When air resistance and other forces are taken into account for real projectile motion, the equations and graphs of motion become more complicated. Again, we can examine $R = -kv$ (linear drag) and $R = -kv^2$ (quadratic drag). Now that we are working in 2 dimensions, the resistance is acting in both horizontal and vertical directions.

With $R = -kv$, the trajectory is no longer a smooth parabola, and the components for acceleration can be found using Newton's second law: $F = m\ddot{x}$.

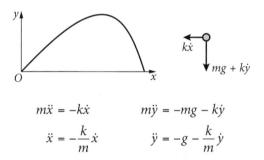

$$m\ddot{x} = -k\dot{x} \qquad m\ddot{y} = -mg - k\dot{y}$$

$$\ddot{x} = -\frac{k}{m}\dot{x} \qquad \ddot{y} = -g - \frac{k}{m}\dot{y}$$

Integration gives equations for velocity and displacement, and eliminating t gives the equation of the trajectory. The theory can be found in your textbook and is quite complex and tedious, so it is unlikely to feature in an HSC exam because solving such problems would be quite time-consuming.

The path of the projectile is not a parabola because the function is more complicated and of the form

$$y = ax + b\ln(1 - cx).$$

Practice set 1

Multiple-choice questions

Solutions start on page 154.

Question 1 ⬤◯◯

An object of mass m is dropped from a very high cliff. The object falls with air resistance (R) equal to mkv^2, where v is its velocity and k is a constant.

Find the terminal velocity of the object.

A $v_T = \dfrac{g}{k}$ **B** $v_T = \sqrt{\dfrac{g}{k}}$ **C** $v_T = \dfrac{k}{g}$ **D** $v_T = \sqrt{\dfrac{k}{g}}$

Question 2 ⬤◯◯

A rock is thrown vertically upwards with a speed of 25 m/s from the edge of a cliff 30 metres above the water.

How long has the rock been in the air before it hits the water below?

A 1 s   **B** 2 s   **C** 3 s   **D** 6 s

Question 3 ⬤◯◯

A particle has an acceleration given by $a = 4x$, where x is the particle's displacement.

The velocity of the particle is −8 m/s when it is 1 m from the origin.

What is the equation for its velocity, v?

A $v = \sqrt{4x^2 + 60}$ **B** $v = -\sqrt{4x^2 + 60}$ **C** $v = 2x^2 - 10$ **D** $v = 2x^2 + 10$

Question 4 ⬤◯◯

The velocity, v m/s, of a particle is given by $v = 12t - 3t^2$, $t \geq 0$, where t is measured in seconds.

Find the displacement of the particle at 4 seconds.

A 0 m   **B** 12 m   **C** 16 m   **D** 32 m

Question 5 ⬤◯◯

A body that is initially at rest has a displacement given by $x = 3t^2 e^{-2t}$, where t is measured in seconds and x in metres.

When is the body at rest a second time?

A 1 s   **B** 3 s   **C** 6 s   **D** 9 s

Question 6 ⬤◯◯

A 5 kg mass is suspended by inextensible string from a horizontal beam as shown.

Find the ratio $T_2 : T_1$.

A $1 : \sqrt{3}$ **B** $\sqrt{3} : 1$

C $2 : 1$ **D** $1 : 2$

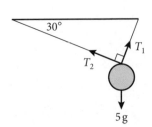

Question 7

A particle of unit mass falls from rest and the resistance is equal to kv^2, where v is its speed and k is a positive constant.

What is the formula for v^2, where x is the distance fallen?

A $v^2 = \frac{g}{k}(1 - e^{-2kx})$　　　**B** $v^2 = \frac{g}{k}(1 + e^{-2kx})$　　　**C** $v^2 = \frac{g}{k}(1 - e^{-2x})$　　　**D** $v^2 = \frac{g}{k}(1 + e^{2x})$

Question 8

Two particles of mass 3 kg and 2 kg are connected by a light string over a smooth pulley.

What is the acceleration of the system when released?

A $3g\,\text{m/s}^2$　　　　　　　　**B** $2g\,\text{m/s}^2$

C $g\,\text{m/s}^2$　　　　　　　　**D** $\frac{g}{5}\,\text{m/s}^2$

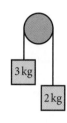

Question 9

Two identical balls, P and Q, are projected horizontally from the top of a building at the same time. The trajectory of each ball is shown.

Which of the following statements are correct?

　　I　They both reach the ground at the same time.

　　II　They both have the same vertical acceleration.

　　III　They have the same final velocity.

A I only　　　　　　　　**B** I and II only

C II and III only　　　　　**D** I, II and III

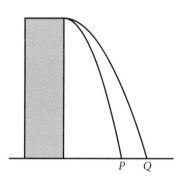

Question 10

A body of mass 5 kg is at rest on a rough horizontal table with coefficient of friction of 0.4.

What is the least horizontal force required to cause the body to move horizontally across the table?

A 2 N　　　　　　　**B** 5 N　　　　　　　**C** 12.5 N　　　　　　　**D** 20 N

Question 11

Two bodies are shown joined by a light inextensible string over a smooth pulley. The system is about to move. F represents the force due to the friction between one of the bodies and the table on which it lies.

What is the coefficient of friction?

A $\dfrac{M}{m}$　　　　　　　　**B** $M - m$

C $\dfrac{m}{M}$　　　　　　　　**D** $m - M$

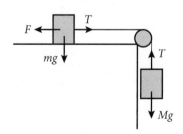

Question 12 ©NESA 2020 HSC EXAM, QUESTION 5 ⬤⬤◯

A particle undergoing simple harmonic motion has a maximum acceleration of $6 \, \text{m/s}^2$ and a maximum velocity of $4 \, \text{m/s}$.

What is the period of the motion?

A π 　　　　　**B** $\dfrac{2\pi}{3}$ 　　　　　**C** 3π 　　　　　**D** $\dfrac{4\pi}{3}$

Question 13 ©NESA 2014 HSC EXAM, QUESTION 9 ⬤⬤◯

A particle is moving along a straight line so that initially its displacement is $x = 1$, its velocity is $v = 2$, and its acceleration is $a = 4$.

Which is a possible equation describing the motion of the particle?

A $v = 2\sin(x-1) + 2$ 　　　　　**B** $v = 2 + 4\log_e x$

C $v^2 = 4(x^2 - 2)$ 　　　　　**D** $v = x^2 + 2x + 4$

Question 14 ⬤⬤⬤

A 40-kilogram box is initially at rest. The box is pulled along the ground with a force of 100 newtons at an angle of $30°$ to the horizontal. The box experiences a resistive force of $0.2R$ newtons, where R is the normal force, as shown in the diagram. (Assume $g = 10 \, \text{m/s}^2$).

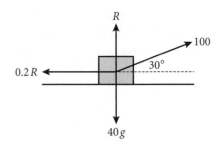

What is the net horizontal force?

A $50\sqrt{3} - 70$ 　　　　　**B** $50\sqrt{3} + 70$

C $70\sqrt{3} - 50$ 　　　　　**D** $70\sqrt{3} + 50$

Question 15 ⬤⬤⬤

A particle moving in a straight line follows the equation $v^2 = -x^2 + 2x + 8$, where x is its displacement from the origin in metres and v is its velocity in m/s.

Given that the motion is simple harmonic, what is the amplitude of this motion?

A $3 \, \text{m}$ 　　　　　**B** $2\pi \, \text{m}$ 　　　　　**C** $8 \, \text{m}$ 　　　　　**D** $9 \, \text{m}$

Question 16 ⬤⬤◯

An object with mass m kilograms falls from a stationary balloon and experiences air resistance during its fall equal to mkv, where $v \, \text{m/s}$ is its speed and k is a positive constant. The height, in metres, of the object above the ground as it falls is x. The acceleration due to gravity is g and the positive direction is taken to be upwards.

Find the equation for the object's acceleration.

A $\ddot{x} = g - kv$ 　　　　　**B** $\ddot{x} = g + kv$

C $\ddot{x} = -g + kv$ 　　　　　**D** $\ddot{x} = -g - kv$

Question 17 ⬤⬤⬤

The acceleration of a particle moving in a straight line with velocity v is given by $\ddot{x} = v$.

Which of the following functions best represents v in terms of x?

A $v = e^x$ 　　　　　**B** $v = \sqrt[3]{3x + 1}$

C $v = 2x + 1$ 　　　　　**D** $v = x + \dfrac{1}{3}$

Question 18 〇〇〇

A particle is moving with simple harmonic motion described by $\ddot{x} = -10(x + 2)$ with amplitude of 5.

Find the velocity of the particle when $x = -3$.

A $v = -2$ **B** $v = 10$ **C** $v = \pm 4\sqrt{15}$ **D** $v = 240$

Question 19 〇〇〇

A particle is undergoing simple harmonic motion and takes 4 seconds to travel between the extremes of the motion at $x = -3$ and $x = 3$.

Which of the following is a possible equation for the velocity of the particle?

A $v = 3\sin\left(\dfrac{\pi}{2}t\right)$ **B** $v = \dfrac{3\pi}{2}\sin\left(\dfrac{\pi}{2}t\right)$

C $v = 3\sin\left(\dfrac{\pi}{4}t\right)$ **D** $v = \dfrac{3\pi}{4}\sin\left(\dfrac{\pi}{4}t\right)$

Question 20 〇〇〇

A body has acceleration given by $a = 4v$, where v is the velocity of the body.

If the velocity of the body is 1 m/s when it is 2 m from the origin, which of the following represents the velocity of the body?

A $v = 3 - 2x$ **B** $v = x - 1$

C $v = 4x - 7$ **D** $v = \dfrac{1}{4}(x + 2)$

Practice set 2

Short-answer questions

Solutions start on page 156.

Question 1 (2 marks) ⬤●●

A particle starts at the origin with velocity 2 m/s and acceleration given by $a = v^2 - v$, where v is the velocity of the particle and $v > 1$.

Find an expression for x, the displacement of the particle, in terms of v.   2 marks

Question 2 (3 marks) ⬤⬤●

Find the coefficient of friction between the mass and the surface if the mass is accelerating down the plane with acceleration of 5 m/s^2. Let $g = 10 \text{ ms}^{-2}$.
(Answer to two decimal places.)   3 marks

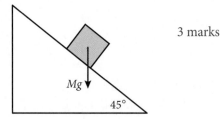

Question 3 (8 marks) ⬤●●

A particle is travelling in a straight line and its position in metres at time t seconds is given by

$$x = 3\sin\left(4t - \frac{\pi}{2}\right).$$

a   What is its initial speed?   2 marks

b   Show that the acceleration is proportional to the displacement.   1 mark

c   State the period of the motion.   1 mark

d   Sketch the displacement-time and the velocity-time graphs.   2 marks

e   What is the distance travelled by the particle in the first $\dfrac{3\pi}{8}$ seconds?   2 marks

Question 4 (6 marks) ⬤⬤●

A particle moves in a straight line such that its position is given by $x = e^{-t}\sin t$.

a   What is the limiting position of the particle?   1 mark

b   Show that $\ddot{x} + 2\dot{x} + 2x = 0$.   3 marks

c   Describe the motion of the particle.   2 marks

Question 5 (2 marks) ⬤●●

A particle of unit mass moves in a straight line against a resistance of $(v + v^2)$, where v is the velocity of this particle. Initially, the particle is at the origin and travelling with speed $U > 0$.

Show that $x = \ln\left|\dfrac{1 + U}{1 + v}\right|$.   2 marks

Question 6 (4 marks) ⬤⬤●

The acceleration of a particle moving in a straight line is $(4t - 5) \text{ m/s}^2$.

The particle starts at the origin with a velocity of 2 m/s.

a   Find the velocity and position of the particle at time t.   2 marks

b   When does the particle first come to rest?   1 mark

c   How often does the particle come to rest?   1 mark

Question 7 (4 marks)

A particle of mass 1 kg moving in a straight line is acted on by a force of $F = 2x^3 + 2x$ newtons, where x is the displacement of the particle in metres.

If the particle is initially 1 metre left of the origin, moving at a velocity of 2 m/s, find the displacement function $x(t)$.

4 marks

Question 8 (6 marks)

An object of mass 2 kg is projected vertically upward with a velocity of 100 m/s. The air resistance acting on the particle is $4v$ newtons, where v is the velocity of the object.

a   Show that the equation of motion is given by $\ddot{x} = -(10 + 2v)$, using $g = 10 \text{ m s}^{-2}$.   1 mark

b   Find the maximum height reached by the object.   3 marks

c   Find the time at which the object reaches its maximum height.   2 marks

Question 9 (5 marks)

A particle is travelling in simple harmonic motion such that $\ddot{x} = -5x$, where x represents the displacement from the origin in metres.

a   Show that $x = a\cos(\sqrt{5}t + \alpha)$ is a possible equation for the displacement of the particle, where a and α are constants and t represents time in seconds.   2 marks

b   The particle is initially observed to have a velocity of $\sqrt{10}$ m/s at a displacement of 4 m from the origin. Find the amplitude (a) of the motion.   2 marks

c   Determine the maximum speed of the particle.   1 mark

Question 10 (4 marks)

A particle is moving in a straight line. At time t seconds, it has displacement x metres from a fixed point O on the line, where $x = 1 + \cos 2t + \sin 2t$.

a   Express x in the form $x = 1 + a\cos(2t - \alpha)$ for constants $a > 0$ and $0 < \alpha < \dfrac{\pi}{2}$.   2 marks

b   Draw the displacement-time graph.   2 marks

Question 11 (5 marks)

A projectile is fired from ground level with an initial velocity of v_0 m/s at an angle of β to the horizontal. The air resistance is directly proportional to the velocity, with k the constant of proportionality. The equations of motion are:

$$x = \frac{v_0 \cos \beta}{k}(1 - e^{-kt}) \qquad y = \frac{10 + kv_0 \sin \beta}{k^2}(1 - e^{-kt}) - \frac{10t}{k}$$

(Do NOT prove these equations.)

A projectile is fired at an angle of 30° to the horizontal, with initial velocity $3\sqrt{3}$ m/s and $k = 0.3$.

a   Find, correct to two decimal places, the time for the projectile to reach its greatest height.   2 marks

b   The projectile lands in a valley below its initial position in 3 seconds. Find, correct to one decimal place, the magnitude of the velocity of the projectile when it lands.   3 marks

Question 12 (6 marks) ●●●

A projectile is fired from the origin with an initial velocity V at an angle of θ to the horizontal.

a   What are the initial vertical and horizontal components of velocity?   2 marks

b   Ignoring air resistance, derive the equations of motion of the projectile and find the   2 marks
equation of the trajectory with y expressed in terms of x.

c   Find a monic quadratic in $\tan\theta$ if the projectile is to land at a point R metres from the   2 marks
launch point.

Question 13 (3 marks) ●●●

A particle initially at the origin moves along a straight line under an acceleration $(4a^2x - 4x^3)$,
where x is the distance from the origin to the particle and a and k are constants.

Its initial velocity is $a^2\sqrt{6}$ in the direction of x increasing.

At what distance from the origin will the particle first come to rest?   3 marks

Question 14 (3 marks) ●●●

Two objects of mass m kg and M kg are joined
together by a light inextensible string, which
goes over a smooth pulley, as in the diagram.
The object with mass m kg is on a smooth
plane inclined at θ to the horizontal.

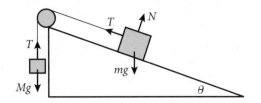

If the ratio $M:m$ is $1:2$ and $g = 10\,\text{m/s}^2$, find the size of θ correct to one decimal place   3 marks
if the object with mass M is accelerating at $2\,\text{m/s}^2$.

Question 15 (3 marks) ●●●

A body of mass 3 kg is dropped from a height and is subjected to a resistance of $\dfrac{v}{5}$ newtons,
where v is the velocity of the body.

Find the velocity of the body, $v(t)$. [Use $g = 10\,\text{m/s}^2$]   3 marks

Question 16 (8 marks) ●●●

A 4 kg projectile is launched at a speed of 500 m/s at an angle of 30° from the horizontal.

It experiences gravity of 40 newtons and air resistance opposite to its direction of motion
of v newtons.

a   Show that $\ddot{x} = -\dfrac{1}{4}\dot{x}$ and $\ddot{y} = -10 - \dfrac{1}{4}\dot{y}$, where x is horizontal displacement and y is   2 marks
vertical displacement.

b   Show that the initial velocities in the horizontal and vertical directions are $250\sqrt{3}$ m/s   1 mark
and 250 m/s, respectively.

c   Show that $\dot{x} = 250\sqrt{3}e^{-\frac{t}{4}}$ and $x = 1000\sqrt{3}(1 - e^{-\frac{t}{4}})$.   4 marks

d   Show that the horizontal displacement of the projectile approaches a constant value.   1 mark

Question 17 (3 marks) ●●●

A 40 kg mass is attached by 2 inextensible wires that are 6 m and 8 m long to the same
horizontal plane at points that are 10 m apart.

Find the tension in the 8 m wire.   3 marks

Question 18 (2 marks) ●●●

The maximum height reached by a projectile is $\dfrac{V^2 \sin^2 \theta}{2g}$, where V is the initial velocity, θ is the angle of projection and g is the gravitational acceleration. (Do NOT prove this.)

Two balls are thrown from the same point at angles of $30°$ and $60°$ to the horizontal and initial velocities V_1 and V_2 respectively.

Find $V_1 : V_2$ if they both reach the same maximum height.   2 marks

Question 19 (6 marks) ©NESA 2013 HSC EXAM, QUESTION 15(d) ●●●

A ball of mass m is projected vertically into the air from the ground with initial velocity u. After reaching the maximum height H it falls back to the ground. While in the air, the ball experiences a resistive force kv^2, where v is the velocity of the ball and k is a constant.

The equation of motion when the ball falls can be written as

$$m\dot{v} = mg - kv \qquad \text{(Do NOT prove this.)}$$

a Show that the terminal velocity v_T of the ball when it falls is $\sqrt{\dfrac{mg}{k}}$.   1 mark

b Show that when the ball goes up, the maximum height H is   3 marks

$$H = \frac{v_T^2}{2g}\ln\left(1 + \frac{u^2}{v_T^2}\right).$$

c When the ball falls from height H it hits the ground with velocity w.

Show that   2 marks

$$\frac{1}{w^2} = \frac{1}{u^2} + \frac{1}{v_T^2}.$$

Question 20 (5 marks) ●●●

A ball is thrown at an angle of $45°$ to the horizontal at $10\,\text{m/s}$ and experiences air resistance proportional to the velocity in both the x and y directions.

The terminal velocity is $2g\,\text{m/s}$.

Show that $x = 10\sqrt{2}\left(1 - e^{-\frac{t}{2}}\right)$ and $y = 2(2g + 5\sqrt{2})\left(1 - e^{-\frac{t}{2}}\right) - 2gt$.   5 marks

PRACTICE SET 2

Practice set 1

Worked solutions

1 B

$$m\ddot{x} = mg - mkv^2$$

For terminal velocity $\ddot{x} = 0$

$$kv^2 = g$$

$$v^2 = \frac{g}{k}$$

So $v_T = \sqrt{\dfrac{g}{k}}$.

2 D

$$\ddot{y} = -g$$
$$\dot{y} = -gt + 25$$
$$y = -\frac{1}{2}gt^2 + 25t + 30$$

Hence, $y = 0$:

$$-5t^2 + 25t + 30 = 0$$
$$t^2 - 5t - 6 = 0$$
$$(t - 6)(t + 1) = 0$$

$$t = -1 \text{ or } 6$$

but $t \geq 0$,

so $t = 6$ s.

3 B

$$\frac{d}{dx}\left(\frac{1}{2}v^2\right) = 4x$$

$$\frac{1}{2}v^2 = 2x^2 + C$$

When $x = 1$, $v = -8$:

$$32 = 2 + C$$
$$C = 30$$

$$v^2 = 4x^2 + 60$$
$$v = -\sqrt{4x^2 + 60}$$

(negative because $v = -8$ when $x = 1$)

4 D

$$v = 12t - 3t^2$$

Displacement $= \displaystyle\int_0^4 12t - 3t^2 \, dt$

$$= \left[6t^2 - t^3\right]_0^4$$

$$= 32 \, \text{m}$$

The particle's movement changes direction when $v = 0$, i.e. $t = 0$ or $t = 4$. Since the direction does not change between $t = 0$ and $t = 4$, the distance travelled is equal to the displacement.

5 A

$$x = 3t^2 e^{-2t}$$

$$v = 6t(e^{-2t}) + 3t^2(-2e^{-2t}) = 0$$
$$6te^{-2t} - 6t^2e^{-2t} = 0$$
$$6te^{-2t}(1 - t) = 0$$
$$t = 0, 1$$

Hence, $t = 1$ s (second time).

6 B

$$\tan 30° = \frac{T_1}{T_2}$$

$$\therefore T_2 : T_1 = \frac{1}{\tan 30°}$$

$$= \frac{1}{\dfrac{1}{\sqrt{3}}}$$

$$= \sqrt{3}$$

$$= \sqrt{3} : 1$$

7 A

$m\ddot{x} = mg - kv^2$, where $m = 1$

$v\dfrac{dv}{dx} = g - kv^2$

$\dfrac{dv}{dx} = \dfrac{g - kv^2}{v}$

$\dfrac{dx}{dv} = \dfrac{v}{g - kv^2}$

$x = -\dfrac{1}{2k}\ln\left|g - kv^2\right| + c$

But $\ddot{x} = g - kv^2 > 0$, so $x = -\dfrac{1}{2k}\ln(g - kv^2) + c$.

When $x = 0$, $v = 0$:

$0 = -\dfrac{1}{2k}\ln g + c$

$c = \dfrac{1}{2k}\ln g$

$\therefore x = -\dfrac{1}{2k}\ln(g - kv^2) + \dfrac{1}{2k}\ln g$

$x = -\dfrac{1}{2k}\ln\left(\dfrac{g - kv^2}{g}\right)$

$-2kx = \ln\left(\dfrac{g - kv^2}{g}\right)$

$e^{-2kx} = \dfrac{g - kv^2}{g}$

$ge^{-2kx} = g - kv^2$

$kv^2 = g - ge^{-2kx}$

$v^2 = \dfrac{g}{k}(1 - e^{-2kx})$

8 D

$(3 + 2)\ddot{x} = 3g - 2g$

$5\ddot{x} = g$

$\ddot{x} = \dfrac{g}{5}$ m/s^2

9 B

Since $\ddot{y} = -g$

$\dot{y} = -gt$

$y = -\dfrac{1}{2}gt^2 + h$

I is correct because both have zero horizontal acceleration and a vertical acceleration of g, so if both travel the same vertical distance then their time of flight is the same.

II is correct because acceleration is due to gravity only.

III is incorrect because the angles at which P and Q hit the ground are different, so their final velocities will be different.

10 D

$F = \mu N$

$N = 5g$

So min $F = 0.4 \times 5g$

$= 20$ N.

11 A

$M\ddot{x} = Mg - T$ and $m\ddot{x} = T - \mu mg$

Adding together:

$(M + m)\ddot{x} = Mg - \mu mg$

$\ddot{x} = \dfrac{(M - \mu m)g}{(M + m)}$

As they are stationary, $\ddot{x} = 0$.

$\therefore M - \mu m = 0$

$\mu = \dfrac{M}{m}$

12 D

$v_{\text{max}} = na = 4$ and $\ddot{x}_{\text{max}} = n^2 a = 6$

Solving simultaneously:

Dividing the 2nd equation by the 1st:

$n = \dfrac{6}{4}$

$T = \dfrac{2\pi}{\frac{6}{4}} = \dfrac{4\pi}{3}$

13 A

Checking each option, you get

$v = 2\sin(x - 1) + 2$,

so $v(1) = 2$

$a(x) = [2\sin(x - 1) + 2] \times 2\cos(x - 1)$

so $a(1) = [0 + 2] \times 2(1) = 4$, as required.

14 A

$F_{\text{net}} = 100\cos 30° - 0.2 \times [40g - 100\sin 30°]$

$= \dfrac{100\sqrt{3}}{2} - 0.2\left(40 \times 10 - 100 \times \dfrac{1}{2}\right)$

$= 50\sqrt{3} - 70$

15 A

$v^2 = -x^2 + 2x + 8$

Complete the square to obtain the form $v^2 = n^2(a^2 - x^2)$:

$v^2 = 9 - (x^2 - 2x + 1) = 1^2\left(3^2 - (x - 1)^2\right)$

Hence, amplitude $a = 3$.

16 C

$m\ddot{x} = -mg + mkv,$

since positive direction is taken upwards.

Hence, $\ddot{x} = -g + kv.$

17 D

$\ddot{x} = \dfrac{d}{dx}\left(\dfrac{1}{2}v^2\right) = v.$

So we could find $\dfrac{d}{dx}\left(\dfrac{1}{2}v^2\right)$ for every option and check which one is equal to v.

But also, $\ddot{x} = v\dfrac{dv}{dx} = v.$

So $\dfrac{dv}{dx} = 1.$

Testing $\dfrac{dv}{dx}$ for every option gives 1 for D only.

18 C

Given $x = -10(x + 2), a = 5, n^2 = 10, c = -2.$

Using $v^2 = 10[25 - (x + 2)^2]$:

Putting $x = -3$:

$v^2 = 10(25 - 1)$

   $= 240$

Hence, $v = \pm\sqrt{240}$

      $= \pm 4\sqrt{15}.$

19 D

Since $a = 3$ and $T = \dfrac{2\pi}{n}$

$8 = \dfrac{2\pi}{n}$

$n = \dfrac{\pi}{4}$

Note: period is 8 seconds,

$\therefore$ let $x = 3\cos\left(\dfrac{\pi}{4}t\right).$

So $v = \dfrac{dx}{dt} = \dfrac{3\pi}{4}\sin\left(\dfrac{\pi}{4}t\right).$

20 C

$\dfrac{dv}{v} = 4dt$

$\ln|v| = 4t + C_1$

$v = \pm e^{4t + C_1} = Ae^{4t}, \text{ where } A = \pm e^{C_1}$

$\dfrac{dx}{dt} = Ae^{4t}$

$x = \dfrac{Ae^{4t}}{4} + C_2$

  $= \dfrac{v}{4} + C_2$

$\therefore 2 = \dfrac{1}{4} + C_2$

$\therefore C_2 = \dfrac{7}{4}$

Hence, $x = \dfrac{v}{4} + \dfrac{7}{4}.$

So $v = 4x - 7.$

Practice set 2

Worked solutions

Question 1

$v\dfrac{dv}{dx} = v^2 - v$

$\dfrac{dv}{v - 1} = dx$

$\ln(v - 1) = x + C \qquad v > 1$

When $x = 0, v = 2$:

$\ln(2 - 1) = 0 + C$

       $0 = C$

So $x = \ln(v - 1).$

Question 2

$M \times 5 = Mg\sin 45° - \mu Mg\cos 45°$

$5 = 10\sin 45° - 10\mu\cos 45°$

$10\mu\cos 45° = 10\sin 45° - 5$

$\mu = \dfrac{10\sin 45° - 5}{10\cos 45°}$

  $= 0.2928\ldots$

  ≈ 0.29

Question 3

a $x = 3\sin\left(4t - \dfrac{\pi}{2}\right)$

$\dot{x} = 12\cos\left(4t - \dfrac{\pi}{2}\right)$

$\dot{x}(0) = 12\cos\left(-\dfrac{\pi}{2}\right) = 0\,\text{m/s}$

b $\dot{x} = 12\cos\left(4t - \dfrac{\pi}{2}\right)$

$\ddot{x} = -48\sin\left(4t - \dfrac{\pi}{2}\right) = -16x$

That is, $\ddot{x}$ is proportional to x.

c Period $T = \dfrac{2\pi}{n}$, $n = 4$

$T = \dfrac{2\pi}{4}$

$= \dfrac{\pi}{2}$

d x-t graph:

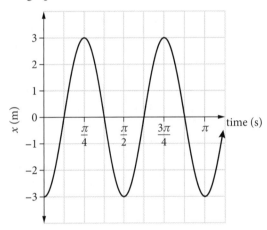

v-t graph:

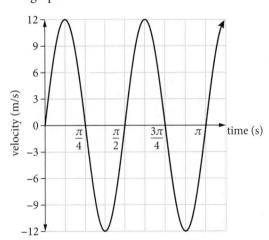

e Distance travelled in first $\dfrac{3\pi}{8}$ seconds:

$t = \dfrac{3\pi}{8}$, $x = 0$.

So the particle has moved from −3 to 3 and back to 0, travelling 9 metres.

Question 4

a $x = e^{-t}\sin t$

$\lim_{t \to \infty} x = 0$

b $\dot{x} = e^{-t}(\cos t - \sin t)$

$\ddot{x} = -2e^{-t}\cos t$

$\therefore \ddot{x} + 2\dot{x} + 2x$

$= -2e^{-t}\cos t + 2e^{-t}(\cos t - \sin t) + 2e^{-t}\sin t$

$= -2e^{-t}\cos t + 2e^{-t}\cos t - 2e^{-t}\sin t + 2e^{-t}\sin t$

$= 0$, as required.

c The particle is oscillating about the origin with decreasing amplitude, approaching 0.

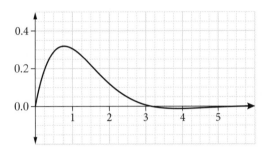

Question 5

$\ddot{x} = -(v + v^2)$

$\dfrac{dv}{1 + v} = -dx$

$\ln|1 + v| = -x + c$

When $x = 0$, $v = U$:

$\ln|1 + U| = 0 + c$

$\therefore \ln|1 + v| = -x + \ln|1 + U|$

$\therefore x = \ln|1 + U| - \ln|1 + v|$

$= \ln\left|\dfrac{1 + U}{1 + v}\right|$, as required.

Question 6

a $\ddot{x} = 4t - 5$

$\dfrac{dv}{dt} = 4t - 5$

$v = 2t^2 - 5t + c$

When $t = 0$, $v = 2$:

$2 = 0 - 0 + c$

$\therefore v = 2t^2 - 5t + 2$

$x = \dfrac{2}{3}t^3 - \dfrac{5}{2}t^2 + 2t + d$

When $t = 0$, $x = 0$:

$0 = 0 - 0 + 0 + d$

So $x = \dfrac{2}{3}t^3 - \dfrac{5}{2}t^2 + 2t$.

b $v = 0$

$$2t^2 - 5t + 2 = 0$$
$$(2t - 1)(t - 2) = 0$$

So $t = \dfrac{1}{2}, 2$.

Particle first comes to rest when $t = \dfrac{1}{2}$ s.

c The particle comes to rest just twice, when $t = \dfrac{1}{2}, 2$.

Question 7

$$\ddot{x} = 2x^3 + 2x$$
$$\frac{d}{dx}\left(\frac{1}{2}v^2\right) = 2x^3 + 2x$$
$$\frac{1}{2}v^2 = \frac{1}{2}x^4 + x^2 + c$$

When $x = -1$, $v = 2$:

$$\frac{1}{2}(2)^2 = \frac{1}{2}(-1)^4 + (-1)^2 + c$$
$$c = \frac{1}{2}$$

So $\dfrac{1}{2}v^2 = \dfrac{1}{2}x^4 + x^2 + \dfrac{1}{2}$

$$v^2 = x^4 + 2x^2 + 1$$
$$= (x^2 + 1)^2.$$

$v = x^2 + 1$ since $v > 0$

$$\frac{dx}{dt} = x^2 + 1$$
$$\frac{dx}{x^2 + 1} = dt$$
$$\tan^{-1}x = t + d$$

When $t = 0$, $x = -1$:

$$\tan^{-1}(-1) = 0 + d$$
$$d = -\frac{\pi}{4}$$

So $\tan^{-1}x = t - \dfrac{\pi}{4}$

$$x = \tan\left(t - \frac{\pi}{4}\right).$$

Question 8

a $2\ddot{x} = -2g - 4v$
So $\ddot{x} = -g - 2v$
$$= -(10 + 2v), \text{ as required.}$$

b $v\dfrac{dv}{dx} = -(10 + 2v)$

$$\frac{v}{10 + 2v}dv = dx$$
$$\left(\frac{1}{2}\frac{(10 + 2v)}{10 + 2v} - \frac{5}{10 + 2v}\right)dv = -dx$$

Integrating:

$$\frac{1}{2}v - \frac{5}{2}\ln|10 + 2v| = -x + c$$

$\ddot{x} = -(10 + 2v) < 0$ going up, so $10 + 2v > 0$.

So $\dfrac{1}{2}v - \dfrac{5}{2}\ln(10 + 2v) = -x + c$

When $t = 0$, $x = 0$, $v = 100$:

$$\frac{1}{2}(100) - \frac{5}{2}\ln[10 + 2(100)] = 0 + c$$
$$50 - \frac{5}{2}\ln 210 = 0 + c$$

So $\dfrac{1}{2}v - \dfrac{5}{2}\ln(10 + 2v) = -x + 50 - \dfrac{5}{2}\ln 210.$

Rearranging for x:

$$x = 50 - \frac{5}{2}\ln(210) - \frac{1}{2}v + \frac{5}{2}\ln(10 + 2v)$$
$$x = 50 - \frac{1}{2}v + \frac{5}{2}\ln\left(\frac{10 + 2v}{210}\right)$$

Maximum height (H) when $v = 0$:

$$H = 50 + \frac{5}{2}\ln\left(\frac{1}{21}\right)$$
$$= 50 - 2.5\ln 21$$

c Time to reach maximum height.

$$\frac{dv}{dt} = -(10 + 2v)$$
$$\frac{dv}{10 + 2v} = -dt$$
$$\frac{1}{2}\ln(10 + 2v) = -t + c$$

When $t = 0$, $v = 100$:

$$\frac{1}{2}\ln(10 + 200) = 0 + c$$
$$\therefore c = \frac{1}{2}\ln 210$$

So $\dfrac{1}{2}\ln(10 + 2v) = -t + \dfrac{1}{2}\ln 210$

$$t = \frac{1}{2}\ln 210 - \frac{1}{2}\ln(10 + 2v)$$
$$= \frac{1}{2}\ln\left(\frac{210}{10 + 2v}\right).$$

When $v = 0$:

$$t = \frac{1}{2}\ln\left(\frac{210}{10 + 0}\right)$$
$$= \frac{1}{2}\ln 21$$

Hence, time to reach maximum height is $\dfrac{1}{2}\ln 21$ seconds.

Question 9

a Let $x = a\cos\left(\sqrt{5}t + \alpha\right)$

$$\dot{x} = -a\sqrt{5}\sin\left(\sqrt{5}t + \alpha\right)$$

$$\ddot{x} = -5a\cos\left(\sqrt{5}t + \alpha\right)$$

So $\ddot{x} = -5x$.

b In simple harmonic motion about the origin:
$v^2 = n^2(a^2 - x^2)$

So $10 = 5(a^2 - 16)$

$a^2 = 18$

$a = 3\sqrt{2}$

c $v^2_{max} = n^2 a^2$

$= 5 \times 18$

$= 90$

$v_{max} = 3\sqrt{10}$ m/s

Question 10

a $x = 1 + \cos 2t + \sin 2t$

$$= 1 + \sqrt{2}\left(\cos 2t \cos\frac{\pi}{4} + \sin 2t \sin\frac{\pi}{4}\right)$$

using the auxiliary angle method

So $x = 1 + \sqrt{2}\cos\left(2t - \frac{\pi}{4}\right)$.

b x-t graph:

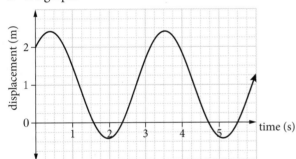

Question 11

a $y = \dfrac{10 + kv_0\sin\beta}{k^2}(1 - e^{-kt}) - \dfrac{10t}{k}$

$\dot{y} = \dfrac{10 + kv_0\sin\beta}{k}(e^{-kt}) - \dfrac{10}{k}$

$\therefore 0 = \dfrac{10 + kv_0\sin\beta}{k}(e^{-kt}) - \dfrac{10}{k},$

$(10 + kv_0\sin\beta)(e^{-kt}) = 10$

That is, $e^{-kt} = \dfrac{10}{10 + kv_0\sin\beta},$

so $t = \dfrac{1}{k}\ln\dfrac{10 + kv_0\sin\beta}{10}.$

Substituting $k = 0.3$, $v_0 = 3\sqrt{3}$, $\beta = 30°$:

$$t = \frac{1}{0.3}\ln\left(\frac{10 + 0.3 \times 3\sqrt{3}\sin 30°}{10}\right)$$

≈ 0.25 s

b $\dot{x} = v_0\cos\beta\, e^{-kt}$, $\dot{y} = \dfrac{10 + kv_0\sin\beta}{k}(e^{-kt}) - \dfrac{10}{k}$

When $t = 3$, $k = 0.3$, $v_0 = 3\sqrt{3}$, $\beta = 30°$:

$$\dot{x} = 3\sqrt{3}\cos 30°\left(e^{-0.3\times 3}\right)$$

≈ 1.8295

$$\dot{y} = \frac{10 + 0.3 \times 3\sqrt{3}\sin 30°}{0.3}\left(e^{-0.3\times 3}\right) - \frac{10}{0.3}$$

≈ -18.7247

Now, $V^2 = 1.8295^2 + (-18.7247)^2 \approx 353.9619$,

Hence, the velocity at which the projectile strikes the ground is $\sqrt{353.9619} \approx 18.8$ m/s.

Question 12

a $\dot{y}(0) = V\sin\theta,$ $\dot{x}(0) = V\cos\theta$

b See table on p.141, Equations of projectile motion, for correct working.

$\ddot{x} = 0$

$\dot{x} = V\cos\theta$

$x = Vt\cos\theta$

$\ddot{y} = -g$

$\dot{y} = -gt + V\sin\theta$

$y = -\dfrac{1}{2}gt^2 + Vt\sin\theta$

Since $t = \dfrac{x}{V\cos\theta}$,

$$y = -\frac{1}{2}g\left(\frac{x}{V\cos\theta}\right)^2 + V\left(\frac{x}{V\cos\theta}\right)\sin\theta$$

$$= -\frac{gx^2}{2V^2\cos^2\theta} + x\left(\frac{\sin\theta}{\cos\theta}\right)$$

$$= \left(-\frac{g}{2V^2\cos^2\theta}\right)x^2 + (\tan\theta)x$$

c For the projectile to land at $y = 0$, $x = R$:

$$0 = \left(-\frac{g}{2V^2\cos^2\theta}\right)R^2 + (\tan\theta)R$$

$$0 = \left(-\frac{g\sec^2\theta}{2V^2}\right)R^2 + (\tan\theta)R$$

$$0 = \left(-\frac{gR^2}{2V^2}\right)(1 + \tan^2\theta) + (\tan\theta)R$$

To make this a monic quadratic in $\tan x$,

multiply each term by $-\dfrac{2V^2}{gR^2}$:

$$1 + \tan^2\theta - \frac{2V^2}{gR^2}(\tan\theta)R = 0$$

$$\tan^2\theta - \frac{2V^2}{gR}(\tan\theta) + 1 = 0$$

Question 13

$$\ddot{x} = 4a^2 x - 4x^3$$

$$\frac{1}{2}v^2 = 2a^2 x^2 - x^4 + C$$

When $x = 0$:

$$v = a^2\sqrt{6},$$

$$\therefore C = 3a^4$$

Hence, $\frac{1}{2}v^2 = 2a^2 x^2 - x^4 + 3a^4$,

$$\therefore v^2 = 4a^2 x^2 - 2x^4 + 6a^4.$$

When $v = 0$:

$$4a^2 x^2 - 2x^4 + 6a^4 = 0$$

which gives $x^4 - 2a^2 x^2 - 3a^4 = 0$.

Factorising:

$$x^4 - 2a^2 x^2 - 3a^4 = (x^2 - 3a^2)(x^2 + a^2)$$

Solving $(x^2 - 3a^2)(x^2 + a^2) = 0$:

$$x^2 = 3a^2$$

$\therefore x = a\sqrt{3}$, $x > 0$ is where particle first comes to rest.

Question 14

For mass M: $M \times 2 = Mg - T$

For mass m: $m \times 2 = T - mg\sin\theta$

Make T the subject of both and equate:

$$Mg - 2M = 2m + mg\sin\theta$$
$$M(10 - 2) = m(2 + 10\sin\theta)$$
$$\frac{M}{m} = \frac{2 + 10\sin\theta}{8} = \frac{1}{2}$$

as $M:m = 1:2$.

Solving for θ:

$$2 + 10\sin\theta = 4,$$

$$\therefore \sin\theta = \frac{2}{10} = \frac{1}{5}$$

$$\theta \approx 11.5°$$

Question 15

$$m\ddot{x} = 3g - \frac{v}{5}$$

$$3\frac{dv}{dt} = 30 - \frac{1}{5}v$$

$$\frac{dv}{dt} = 10 - \frac{1}{15}v$$

$$= \frac{150 - v}{15}$$

Integrating:

$$\int \frac{15\,dv}{150 - v} = \int dt$$

$$-15\ln|150 - v| = t + C$$

But $\ddot{x} = \frac{150 - v}{15} > 0$, so $150 - v > 0$.

So $-15\ln(150 - v) = t + C$.

When $t = 0$, $v = 0$:

so $C = -15\ln 150$

$$t = 15\ln\left(\frac{150}{150 - v}\right),$$

which rearranges to give

$$v = 150\left(1 - e^{-\frac{t}{15}}\right).$$

Question 16

a $4\ddot{x} = -\dot{x}$ and $4\ddot{y} = -40 - \dot{y}$

$$\therefore \ddot{x} = -\frac{1}{4}\dot{x} \qquad \therefore \ddot{y} = -10 - \frac{1}{4}\dot{y}, \text{ as required.}$$

b $\dot{x}(0) = 500\cos 30°$
$$= 250\sqrt{3} \text{ m/s}$$

$\dot{y}(0) = 500\sin 30°$
$$= 250 \text{ m/s}$$

c   In the x-direction:

$$\frac{d\dot{x}}{\dot{x}} = -\frac{1}{4}dt$$

$$\ln(\dot{x}) = -\frac{1}{4}t + C$$

Given $\dot{x}(0) = 250\sqrt{3}$

$$C = \ln\left(250\sqrt{3}\right)$$

Hence, $\ln(\dot{x}) = -\frac{1}{4}t + \ln\left(250\sqrt{3}\right)$

$$\dot{x} = 250\sqrt{3}\,e^{-\frac{1}{4}t}.$$

So $x = -1000\sqrt{3}\,e^{-\frac{1}{4}t} + D$.

When $t = 0$, $x = 0$, $D = 1000\sqrt{3}$.

Hence, $x = -1000\sqrt{3}\,e^{-\frac{1}{4}t} + 1000\sqrt{3}$

$$\therefore x = 1000\sqrt{3}(1 - e^{-\frac{1}{4}t}).$$

d   Range of particle $= \lim_{t \to \infty} x = 1000\sqrt{3}$

Question 17

Using the diagram:

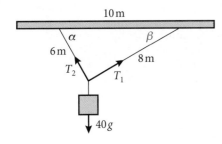

Since the triangle is $6:8:10$, it is right-angled.

Resolving forces:

$$T_1 \sin \beta + T_2 \sin \alpha = 40g \qquad [1]$$
$$T_1 \cos \beta = T_2 \cos \alpha \qquad [2]$$
$$T_2 = \frac{\cos \beta}{\cos \alpha} T_1 = \frac{4}{3} T_1$$

Substitute into [1]:

$$T_1 \sin \beta + \frac{4}{3} T_1 \sin \alpha = 40g$$

$$T_1 = \frac{40g}{\sin \beta + \frac{4}{3} \sin \alpha}$$

$$= \frac{400}{\frac{3}{5} + \frac{4}{3} \times \frac{4}{5}},$$

$$= 240\,\text{N}$$

Question 18

$$\text{Maximum height} = \frac{V_1^2 \sin^2 30°}{2g} = \frac{V_2^2 \sin^2 60°}{2g}$$

$$V_1^2 \left(\frac{1}{2}\right)^2 = V_2^2 \left(\frac{\sqrt{3}}{2}\right)^2$$

$$V_1^2 \left(\frac{1}{4}\right) = V_2^2 \left(\frac{3}{4}\right)$$

$$\frac{V_1^2}{V_2^2} = 3$$

$$\frac{V_1}{V_2} = \sqrt{3}$$

$$V_1 : V_2 = \sqrt{3} : 1$$

Note: There are other methods to show this by equating expressions for the maximum height or time of flight.

Question 19

a $\quad m\dot{v} = mg - kv^2$

For terminal velocity $\ddot{x} = 0$:

$$mg - kv^2 = 0$$
$$v_T^2 = \frac{mg}{k}$$
$$v_T = \sqrt{\frac{mg}{k}}$$

b $\qquad \dot{v} = v\frac{dv}{dx} = -g - \frac{kv^2}{m}$

$$-\frac{v}{g + \frac{kv^2}{m}} = \frac{dx}{dv}$$

$$x = -\frac{1}{2}\frac{m}{k} \ln\left(g + \frac{kv^2}{m}\right) + c$$

When the ball goes up, $\ddot{x} = -g - \frac{kv^2}{m} < 0$,

so $g + \frac{kv^2}{m} > 0$.

When $x = 0$, $v = u$:

$$0 = -\frac{m}{2k} \ln\left(g + \frac{ku^2}{m}\right) + c$$

$$c = \frac{m}{2k} \ln\left(g + \frac{ku^2}{m}\right)$$

$$\therefore x = -\frac{m}{2k} \ln\left(g + \frac{kv^2}{m}\right) + \frac{m}{2k} \ln\left(g + \frac{ku^2}{m}\right)$$

$$= \frac{m}{2k} \left[\ln\left(\frac{mg + ku^2}{m}\right) - \ln\left(\frac{mg + kv^2}{m}\right)\right]$$

$$= \frac{m}{2k} \ln\left(\frac{\frac{mg + ku^2}{m}}{\frac{mg + kv^2}{m}}\right)$$

$$= \frac{m}{2k} \ln\left(\frac{mg + ku^2}{mg + kv^2}\right)$$

When $v = 0$, $x = H$:

$$H = \frac{m}{2k} \ln\left(\frac{mg + ku^2}{mg}\right)$$

$$= \frac{m}{2k} \ln\left(1 + \frac{ku^2}{mg}\right)$$

But $v_T = \sqrt{\frac{mg}{k}}$ from part **a**.

$$\therefore v_T^2 = \frac{mg}{k} \text{ or } \frac{m}{k} = \frac{v_T^2}{g}$$

$$\therefore H = \frac{v_T^2}{2g} \ln\left(1 + \frac{u^2}{v_T^2}\right), \text{ as required.}$$

c For the motion downward:

$$m\ddot{x} = mg - kv^2$$

$$\ddot{x} = v\frac{dv}{dx} = g - \frac{kv^2}{m}$$

$$\frac{v}{g - \frac{kv^2}{m}} = \frac{dx}{dv}$$

$$x = -\frac{1}{2}\frac{m}{k}\ln\left(g - \frac{kv^2}{m}\right) + d$$

When the ball goes down, $\ddot{x} = g - \dfrac{kv^2}{m} > 0$.

When $x = 0$, $v = 0$:

$$0 = -\frac{m}{2k}\ln g + d$$

$$d = \frac{m}{2k}\ln g$$

$$\therefore x = -\frac{m}{2k}\ln\left(g - \frac{kv^2}{m}\right) + \frac{m}{2k}\ln g$$

$$= \frac{m}{2k}\left[\ln g - \ln\left(\frac{mg - kv^2}{m}\right)\right]$$

$$= \frac{m}{2k}\ln\left(\frac{g}{\frac{mg - kv^2}{m}}\right)$$

$$= \frac{m}{2k}\ln\left(\frac{mg}{mg - kv^2}\right)$$

When $x = H$, $v = w$:

$$H = \frac{m}{2k}\ln\left(\frac{mg}{mg - kw^2}\right)$$

But $v_T^2 = \dfrac{mg}{k}$ or $\dfrac{m}{k} = \dfrac{v_T^2}{g}$.

$$H = \frac{v_T^2}{2g}\ln\left(\frac{1}{1 - \frac{kw^2}{mg}}\right)$$

$$= \frac{v_T^2}{2g}\ln\left(\frac{1}{1 - \frac{w^2}{v_T^2}}\right).$$

But $H = \dfrac{v_T^2}{2g}\ln\left(1 + \dfrac{u^2}{v_T^2}\right)$ from part **b**.

$$\therefore \frac{v_T^2}{2g}\ln\left(1 + \frac{u^2}{v_T^2}\right) = \frac{v_T^2}{2g}\ln\left(\frac{1}{1 - \frac{w^2}{v_T^2}}\right)$$

$$\therefore 1 + \frac{u^2}{v_T^2} = \frac{1}{1 - \frac{w^2}{v_T^2}}$$

$$\therefore \frac{v_T^2 + u^2}{v_T^2} = \frac{v_T^2}{v_T^2 - w^2}$$

$$(v_T^2 - w^2)(v_T^2 + u^2) = v_T^4$$

$$v_T^4 + v_T^2 u^2 - w^2 v_T^2 - w^2 u^2 = v_T^4$$

$$v_T^2 u^2 = w^2 v_T^2 + w^2 u^2$$

Dividing both sides by $v_T^2 w^2 u^2$:

$$\frac{1}{w^2} = \frac{1}{u^2} + \frac{1}{v_T^2}, \text{ as required.}$$

Question 20

$\ddot{y} = -g - k\dot{y}$ (going up)

$v_T = -2g$ (negative value because terminal velocity is down)

For terminal velocity: $\ddot{y} = 0$,

$$0 = -g - k(-2g)$$

$$-2kg = -g$$

$$k = \frac{1}{2}$$

Horizontal motion

$$\ddot{x} = -k\dot{x}$$

$$\frac{d\dot{x}}{\dot{x}} = -\frac{1}{2}dt$$

$$\ln\dot{x} = -\frac{1}{2}t + C \quad (\ddot{x} > 0, \text{ so } \dot{x} > 0)$$

When $t = 0$, $\dot{x} = 10\cos 45°$:

$$C = \ln(10\cos 45°) = \ln\left(5\sqrt{2}\right)$$

$$\ln\dot{x} = -\frac{1}{2}t + \ln\left(5\sqrt{2}\right)$$

$$\dot{x} = 5\sqrt{2}e^{-\frac{1}{2}t}$$

$$dx = 5\sqrt{2}e^{-\frac{t}{2}}dt$$

$$x = -10\sqrt{2}e^{-\frac{t}{2}} + D$$

When $t = 0$, $x = 0$: $D = 10\sqrt{2}$

$$x = -10\sqrt{2}e^{-\frac{t}{2}} + 10\sqrt{2}$$

$$x = 10\sqrt{2}\left(1 - e^{-\frac{t}{2}}\right)$$

Vertical motion

$$\ddot{y} = -g - k\dot{y} \quad (\ddot{y} < 0, \text{ so } g + k\dot{y} > 0)$$

$$\frac{d\dot{y}}{g + \frac{1}{2}\dot{y}} = -dt$$

$$2\ln\left(g + \frac{1}{2}\dot{y}\right) = -t + E$$

When $t = 0$, $\dot{y} = 10\sin 45° = 5\sqrt{2}$:

$$E = 2\ln\left(g + \frac{5}{2}\sqrt{2}\right)$$

$$2\ln\left(g + \frac{1}{2}\dot{y}\right) = -t + 2\ln\left(g + \frac{5}{2}\sqrt{2}\right)$$

$$t = 2\ln\left(g + \frac{5}{2}\sqrt{2}\right) - 2\ln\left(g + \frac{1}{2}\dot{y}\right)$$

$$\frac{t}{2} = \ln\left(\frac{g + \frac{5}{2}\sqrt{2}}{g + \frac{1}{2}\dot{y}}\right)$$

$$\left(g + \frac{5}{2}\sqrt{2}\right)e^{-\frac{t}{2}} = g + \frac{1}{2}\dot{y}$$

$$\dot{y} = \left(2g + 5\sqrt{2}\right)e^{-\frac{t}{2}} - 2g$$

Now, integrating:

$$y = \frac{\left(2g + 5\sqrt{2}\right)e^{-\frac{t}{2}}}{-\frac{1}{2}} - 2gt + F$$

$$= -2(2g + 5\sqrt{2})e^{-\frac{t}{2}} - 2gt + F.$$

When $t = 0$, $y = 0$:

$$\therefore F = 2\left(2g + 5\sqrt{2}\right)$$

Hence,

$$y = -2\left(2g + 5\sqrt{2}\right)e^{-\frac{t}{2}} - 2gt + 2\left(2g + 5\sqrt{2}\right)$$

$$= 2(2g + 5\sqrt{2})\left(1 - e^{-\frac{t}{2}}\right) - 2gt.$$

HSC exam topic grid (2011–2020)

This table shows the coverage of this topic in past HSC exams by question number. The past exams can be downloaded from the NESA website (www.educationstandards.nsw.edu.au) by selecting 'Year 11 – Year 12', 'HSC exam papers'. NESA marking feedback and guidelines can also be found there.

The new Mathematics Extension 2 course was first examined in 2020. For exams before 2020, select 'Year 11 – Year 12', 'Resources archive', 'HSC exam papers archive'.

Before 2020, simple harmonic motion and velocity/acceleration as functions of x were in the Mathematics Extension 1 course.

	Simple harmonic motion, $v = f(x)$	Modelling motion	Resisted motion	Projectile motion
2011	3(a)*		6(a)	
2012	6*, 13(c)*		13(a)	14(b)*
2013	12(e)*		**15(d)**	13(c)*
2014	7*, **9**, 12(a)*, 12(c)*		14(c)	14(a)*
2015	9*, 13(a)*, 14(b)*		15(a)	14(a)*
2016	**13(a)***	15(b)		13(b)*
2017	12(d)*, 13(a)*		13(c)	13(c)*
2018	7*, 10*		**14(b)**	13(c)*
2019	5*, 12(b)*		**14(b)**	13(c), 13(d)*
2020 new course	**5**, 11(c), **13(a)**	**12(a)**, 16(a)	**14(b)**	**12(b)**

Questions in **bold** can be found in this chapter.

* Mathematics Extension 1 exam

HSC exam reference sheet

Mathematics Advanced, Extension 1 and Extension 2

© NSW Education Standards Authority

Note: Unlike the actual HSC exam reference sheet, this sheet indicates which formulas are Mathematics Extension 1 and Mathematics Extension 2.

Measurement

Length

$$l = \frac{\theta}{360} \times 2\pi r$$

Area

$$A = \frac{\theta}{360} \times \pi r^2$$

$$A = \frac{h}{2}(a + b)$$

Surface area

$$A = 2\pi r^2 + 2\pi rh$$

$$A = 4\pi r^2$$

Volume

$$V = \frac{1}{3}Ah$$

$$V = \frac{4}{3}\pi r^3$$

Functions

$$x = \frac{-b \pm \sqrt{b^2 - 4ac}}{2a}$$

For $ax^3 + bx^2 + cx + d = 0$:*   *EXT1

$$\alpha + \beta + \gamma = -\frac{b}{a}$$

$$\alpha\beta + \alpha\gamma + \beta\gamma = \frac{c}{a}$$

$$\text{and } \alpha\beta\gamma = -\frac{d}{a}$$

Relations

$$(x - h)^2 + (y - k)^2 = r^2$$

Financial Mathematics

$$A = P(1 + r)^n$$

Sequences and series

$$T_n = a + (n - 1)d$$

$$S_n = \frac{n}{2}[2a + (n - 1)d] = \frac{n}{2}(a + l)$$

$$T_n = ar^{n-1}$$

$$S_n = \frac{a(1 - r^n)}{1 - r} = \frac{a(r^n - 1)}{r - 1}, r \neq 1$$

$$S = \frac{a}{1 - r}, |r| < 1$$

Logarithmic and Exponential Functions

$$\log_a a^x = x = a^{\log_a x}$$

$$\log_a x = \frac{\log_b x}{\log_b a}$$

$$a^x = e^{x \ln a}$$

Trigonometric Functions

$\sin A = \dfrac{\text{opp}}{\text{hyp}}$, $\cos A = \dfrac{\text{adj}}{\text{hyp}}$, $\tan A = \dfrac{\text{opp}}{\text{adj}}$

$A = \dfrac{1}{2} ab \sin C$

$\dfrac{a}{\sin A} = \dfrac{b}{\sin B} = \dfrac{c}{\sin C}$

$c^2 = a^2 + b^2 - 2ab \cos C$

$\cos C = \dfrac{a^2 + b^2 - c^2}{2ab}$

$l = r\theta$

$A = \dfrac{1}{2} r^2 \theta$

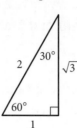

Trigonometric identities

$\sec A = \dfrac{1}{\cos A}$, $\cos A \neq 0$

$\operatorname{cosec} A = \dfrac{1}{\sin A}$, $\sin A \neq 0$

$\cot A = \dfrac{\cos A}{\sin A}$, $\sin A \neq 0$

$\cos^2 x + \sin^2 x = 1$

Compound angles*

$\sin(A + B) = \sin A \cos B + \cos A \sin B$

$\cos(A + B) = \cos A \cos B - \sin A \sin B$

$\tan(A + B) = \dfrac{\tan A + \tan B}{1 - \tan A \tan B}$

If $t = \tan \dfrac{A}{2}$, then $\sin A = \dfrac{2t}{1 + t^2}$

$\cos A = \dfrac{1 - t^2}{1 + t^2}$

$\tan A = \dfrac{2t}{1 - t^2}$

$\cos A \cos B = \dfrac{1}{2}\left[\cos(A - B) + \cos(A + B)\right]$

$\sin A \sin B = \dfrac{1}{2}\left[\cos(A - B) - \cos(A + B)\right]$

$\sin A \cos B = \dfrac{1}{2}\left[\sin(A + B) + \sin(A - B)\right]$

$\cos A \sin B = \dfrac{1}{2}\left[\sin(A + B) - \sin(A - B)\right]$

$\sin^2 nx = \dfrac{1}{2}(1 - \cos 2nx)$

$\cos^2 nx = \dfrac{1}{2}(1 + \cos 2nx)$

Statistical Analysis

$z = \dfrac{x - \mu}{\sigma}$

An outlier is a score less than $Q_1 - 1.5 \times \text{IQR}$
or
more than $Q_3 + 1.5 \times \text{IQR}$

Normal distribution

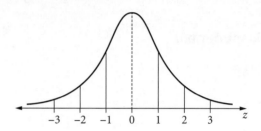

- approximately 68% of scores have z-scores between -1 and 1
- approximately 95% of scores have z-scores between -2 and 2
- approximately 99.7% of scores have z-scores between -3 and 3

Discrete random variables

$E(X) = \mu$

$\operatorname{Var}(X) = E\left[(X - \mu)^2\right] = E(X^2) - \mu^2$

Probability

$P(A \cap B) = P(A)P(B)$

$P(A \cup B) = P(A) + P(B) - P(A \cap B)$

$P(A|B) = \dfrac{P(A \cap B)}{P(B)}$, $P(B) \neq 0$

Continuous random variables

$P(X \leq r) = \displaystyle\int_a^r f(x)\,dx$

$P(a < X < b) = \displaystyle\int_a^b f(x)\,dx$

Binomial distribution*

$P(X = r) = {}^nC_r\, p^r (1 - p)^{n-r}$

$X \sim \text{Bin}(n, p)$
$\Rightarrow P(X = x)$

$= \dbinom{n}{x} p^x (1 - p)^{n-x}$, $x = 0, 1, \ldots, n$

$E(X) = np$

$\operatorname{Var}(X) = np(1 - p)$

*EXT1

Differential Calculus

Function	Derivative
$y = f(x)^n$	$\dfrac{dy}{dx} = nf'(x)[f(x)]^{n-1}$
$y = uv$	$\dfrac{dy}{dx} = u\dfrac{dv}{dx} + v\dfrac{du}{dx}$
$y = g(u)$ where $u = f(x)$	$\dfrac{dy}{dx} = \dfrac{dy}{du} \times \dfrac{du}{dx}$
$y = \dfrac{u}{v}$	$\dfrac{dy}{dx} = \dfrac{v\dfrac{du}{dx} - u\dfrac{dv}{dx}}{v^2}$
$y = \sin f(x)$	$\dfrac{dy}{dx} = f'(x)\cos f(x)$
$y = \cos f(x)$	$\dfrac{dy}{dx} = -f'(x)\sin f(x)$
$y = \tan f(x)$	$\dfrac{dy}{dx} = f'(x)\sec^2 f(x)$
$y = e^{f(x)}$	$\dfrac{dy}{dx} = f'(x)e^{f(x)}$
$y = \ln f(x)$	$\dfrac{dy}{dx} = \dfrac{f'(x)}{f(x)}$
$y = a^{f(x)}$	$\dfrac{dy}{dx} = (\ln a)f'(x)a^{f(x)}$
$y = \log_a f(x)$	$\dfrac{dy}{dx} = \dfrac{f'(x)}{(\ln a)f(x)}$
$y = \sin^{-1} f(x)$	$\dfrac{dy}{dx} = \dfrac{f'(x)}{\sqrt{1 - [f(x)]^2}}$ *
$y = \cos^{-1} f(x)$	$\dfrac{dy}{dx} = -\dfrac{f'(x)}{\sqrt{1 - [f(x)]^2}}$ *
$y = \tan^{-1} f(x)$	$\dfrac{dy}{dx} = \dfrac{f'(x)}{1 + [f(x)]^2}$ *

Integral Calculus

$$\int f'(x)[f(x)]^n dx = \frac{1}{n+1}[f(x)]^{n+1} + c$$
$$\text{where } n \neq -1$$

$$\int f'(x)\sin f(x)\, dx = -\cos f(x) + c$$

$$\int f'(x)\cos f(x)\, dx = \sin f(x) + c$$

$$\int f'(x)\sec^2 f(x)\, dx = \tan f(x) + c$$

$$\int f'(x)e^{f(x)}\, dx = e^{f(x)} + c$$

$$\int \frac{f'(x)}{f(x)}\, dx = \ln|f(x)| + c$$

$$\int f'(x)a^{f(x)}\, dx = \frac{a^{f(x)}}{\ln a} + c$$

$$\int \frac{f'(x)}{\sqrt{a^2 - [f(x)]^2}}\, dx = \sin^{-1}\frac{f(x)}{a} + c \ *$$

$$\int \frac{f'(x)}{a^2 + [f(x)]^2}\, dx = \frac{1}{a}\tan^{-1}\frac{f(x)}{a} + c \ *$$

$$\int u\frac{dv}{dx}\, dx = uv - \int v\frac{du}{dx}\, dx \ **$$

$$\int_a^b f(x)\, dx$$
$$\approx \frac{b-a}{2n}\{f(a) + f(b) + 2[f(x_1) + \cdots + f(x_{n-1})]\}$$
$$\text{where } a = x_0 \text{ and } b = x_n$$

*EXT1, **EXT2

9780170459266

Combinatorics*

$$^nP_r = \frac{n!}{(n-r)!}$$

$$\cdot\binom{n}{r} = \,^nC_r = \frac{n!}{r!(n-r)!}$$

$$(x+a)^n = x^n + \binom{n}{1}x^{n-1}a + \cdots + \binom{n}{r}x^{n-r}a^r + \cdots + a^n$$

Vectors*

$$\left|\underset{\sim}{u}\right| = \left|x\underset{\sim}{i} + y\underset{\sim}{j}\right| = \sqrt{x^2 + y^2}$$

$$\underset{\sim}{u} \cdot \underset{\sim}{v} = \left|\underset{\sim}{u}\right|\left|\underset{\sim}{v}\right|\cos\theta = x_1x_2 + y_1y_2,$$
where $\underset{\sim}{u} = x_1\underset{\sim}{i} + y_1\underset{\sim}{j}$
and $\underset{\sim}{v} = x_2\underset{\sim}{i} + y_2\underset{\sim}{j}$

$$\underset{\sim}{r} = \underset{\sim}{a} + \lambda\underset{\sim}{b}\,{}^{**}$$

Complex Numbers**

$$z = a + ib = r(\cos\theta + i\sin\theta)$$
$$= re^{i\theta}$$

$$\left[r(\cos\theta + i\sin\theta)\right]^n = r^n(\cos n\theta + i\sin n\theta)$$
$$= r^n e^{in\theta}$$

Mechanics**

$$\frac{d^2x}{dt^2} = \frac{dv}{dt} = v\frac{dv}{dx} = \frac{d}{dx}\left(\frac{1}{2}v^2\right)$$

$$x = a\cos(nt + \alpha) + c$$

$$x = a\sin(nt + \alpha) + c$$

$$\ddot{x} = -n^2(x - c)$$

*EXT1, **EXT2

Index